Treasures from the Earth:
The World of Rocks & Minerals

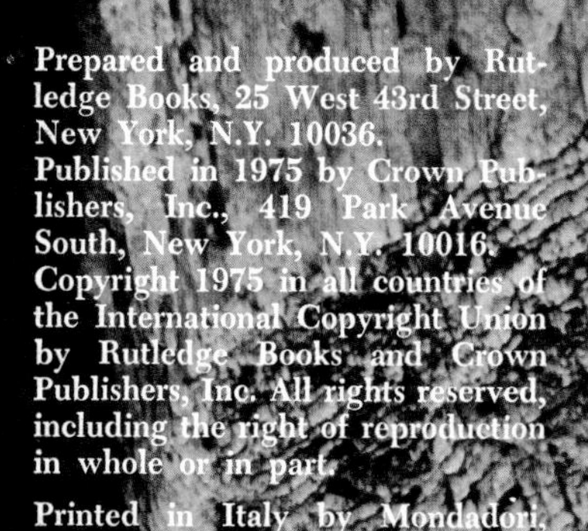

Prepared and produced by Rutledge Books, 25 West 43rd Street, New York, N.Y. 10036.
Published in 1975 by Crown Publishers, Inc., 419 Park Avenue South, New York, N.Y. 10016.

Printed in Italy by Mondadori, Verona.

First printing.

Library of Congress Cataloging in Publication Data

Shaub, Benjamin Martin.
Treasures from the earth: The World of Rocks and Minerals

Bibliography: p. 216
Includes index.
1. Rocks. 2. Mineralogy. I. Title.
QE432.S52 1975 552 75-14313
ISBN 0-517-52347-7

Treasures from the Earth: The World of Rocks & Minerals

by Benjamin M. Shaub

Illustrations by Benjamin M. Shaub and Mary S. Shaub

Consultant:
Joel E. Arem

A Rutledge Book
Crown Publishers, Inc. New York

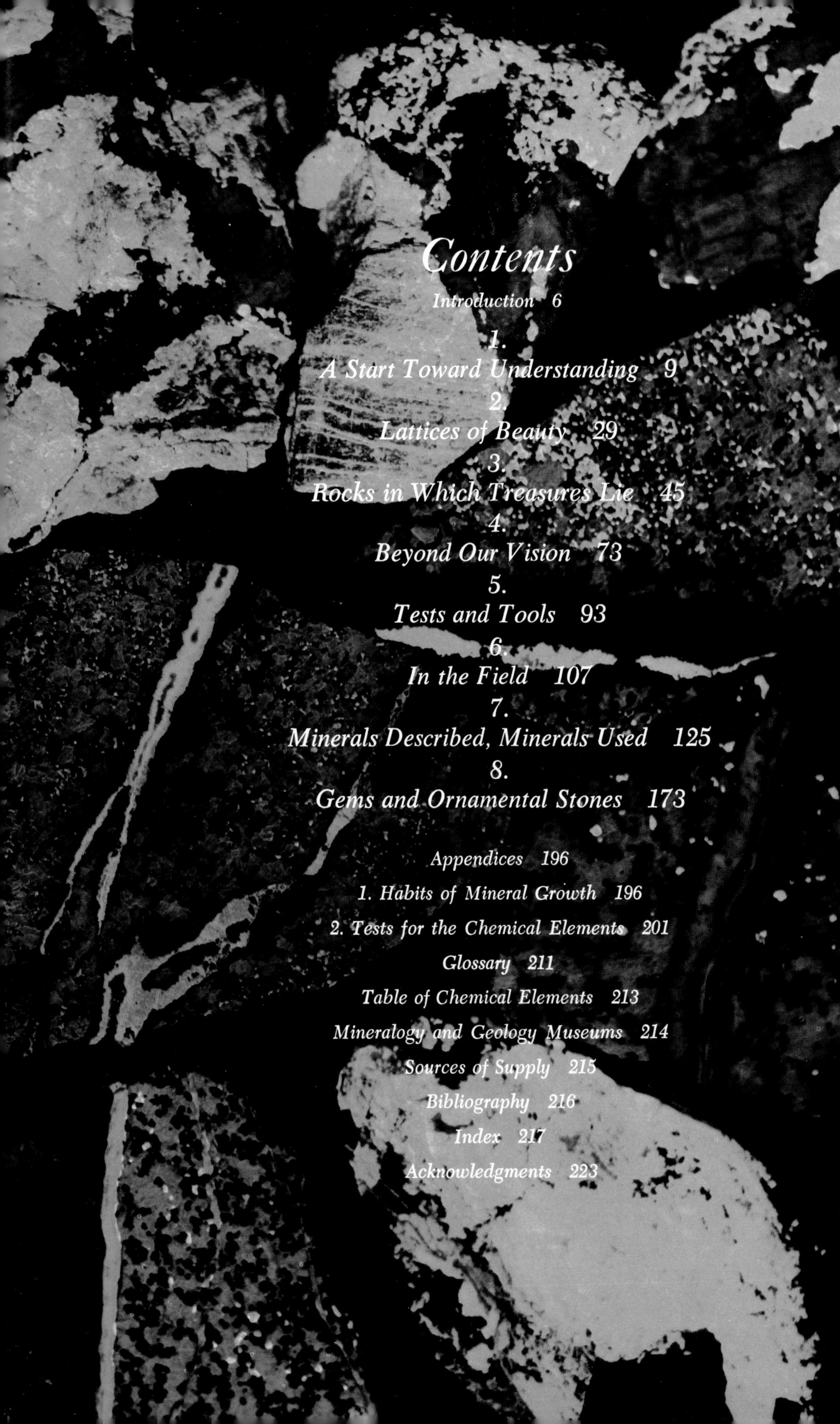

Contents

Introduction

Human civilization, when viewed against the total history of life on earth, is extremely brief. Yet in the past few centuries, human beings have succeeded in nearly exhausting the abundant resources of mineral raw materials once considered inexhaustible. The next few generations will be faced with increasing shortages of vital fuels and metals. Knowledge of the origin of these precious resources will help our society use them more wisely. An understanding of rocks and minerals will also lay the groundwork for judicious conservation policies and the effective use and reuse of scarce natural materials.

The study and appreciation of minerals have advanced simultaneously in the twentieth century. The science of mineralogy has truly come of age, and has contributed greatly to the advancement of our technological society, while a strong amateur movement has made the collection and study of rocks and minerals one of the fastest-growing hobbies in the world. This has enabled the layperson to acquire a greater appreciation of both the value of mineral raw materials in our economy and the beauty and value of fine mineral specimens.

The amateur mineralogist is interested in learning as much as he can about minerals—where they are found, how they form, and how they may be identified. This is the scientific or technical aspect of the mineral hobby. But at the same time, the collector is conscious of the importance of minerals as resources. He wants to know how minerals are used, and how they relate to his life and future. This book addresses itself to those needs.

The classification of minerals according to chemistry has become an accepted practice among mineralogists around the world. In this book, however, minerals are presented for their intrinsic interest and for their usefulness to society. The classification used here reflects this approach; it is based on mineral origins, mineral forms, on the metallic content of those minerals that can be considered ores, and on the uses of nonore minerals.

Many species of minerals have no major uses in industry, but because of their crystallization, color, or beauty they have become popular among mineral collectors. Amateurs can learn to recognize at sight many of the so-called common minerals. But many species are not readily recognized. In such a case, a number of easy tests or experiments can be made to check the properties of an "unknown" mineral specimen; from the results the material can usually be identified. These tests should ideally be based on the use of inexpensive devices accessible to the amateur. The one tool of greatest simplicity and usefulness in this regard—one that has been used for years, and whose technique has been developed to a fine art—is the blowpipe.

Since blowpipe analysis provides a great deal of vital information about a mineral, and since its use can be easily learned and applied, it has been given great emphasis here. Other methods of analysis discussed are those based on simple physical properties and on the forms of crystals. An experienced collector, with practice and after handling many specimens, may be able to recognize at sight more than one hundred mineral species, based on such readily apparent features as color, crystal shape, luster, and specific gravity.

Sections of this book are devoted to the origin and occurrence of minerals and gems, others to a study of various rock types and the kinds of minerals and gemstones that originate in them. While many rocks produce a variety of minerals, large concentrations of certain minerals that can be considered exploitable deposits have definite relationships to different rocks. The identity of a mineral is easier to determine if its host rock is known.

There are many ways in which the beginning mineral collector can get started in his new hobby. Local mineral and gem clubs abound in the United States, and knowledgeable friends who can help with identification problems are easily acquired. The clubs also sponsor field trips to mineral localities. These are important, not only because they provide the opportunity to collect specimens, but also because they give the collector an opportunity to learn about the localities themselves and see the minerals in their geologic settings. The ardent, industrious collector may occasionally find a specimen that resists all his attempts at identification—but he can take comfort from knowing that this is how many new mineral species are discovered. When in doubt, the advice of professionals should be sought. Museums or university geology departments are usually willing to be helpful.

In the decade of the 1970s, the world has become aware of the truly limited nature of its natural resources. The first major shortage has involved petroleum—but this is merely a taste of what we can expect in the future, for one after another of the mineral resources that we take for granted. The time has never been better for people to become more aware of the nature of these resources, of how precious, important, and limited they are, and how essential are methods of conservation and recycling. Unless the public gains this knowledge and acts upon it, shortages and crises may become chronic in our society. The development of such awareness is a major objective of this book.

1. *A Start Toward Understanding*

Mineralogy is the study of the naturally occurring, solid chemical materials of the earth, the planets, and other celestial bodies. These chemicals, called minerals, are used by civilization as a source of many raw materials, including metals. Mineralogy helps us decipher the origin, chemistry, occurrence, and properties of minerals.

Those who explore the earth's crust for economic raw materials must know something of the body of available information about minerals, and they must especially be familiar with the common minerals they may encounter in their work. Mineral seekers should know what chemical elements these minerals contain and be able to distinguish them on the basis of their physical and chemical properties. There are many records of valuable mineral deposits that were initially bypassed because the minerals were improperly identified.

WHAT IS A MINERAL?

The mineral kingdom is generally thought of as including all naturally occurring inorganic materials. Minerals can be identified on the basis of their characteristic properties. The study of minerals has established these properties and linked them to the presence of a fairly definite chemical composition, which may vary to some degree, but only within defined limits. Another requirement of a mineral species is the presence of a definite crystalline structure—that is, the chemical elements of each species must be arranged in a fixed geometric pattern.

A mineral may be a chemical element or compound and still fit these requirements. On the other hand, there exist some naturally occurring inorganic materials that have no fixed crystalline structure, such as opal and natural glass. These materials, though included in the mineral kingdom, are not strictly minerals; they are known rather as mineraloids. In addition, substances such as coal, petroleum, and related compounds cannot be considered minerals, because they lack both a definite composition and a crystalline structure. Nevertheless, they are included in statistics relating to mineral resources and are of the utmost importance in our economy.

Minerals are natural substances from which all materials useful to humankind are derived, either directly or indirectly. Plants generate the oxygen we breathe and are also the basis for nearly all animal foods. Plants require foods, also, and these include phosphates, sulfates, nitrates, and potash, all mineral-derived substances. Minerals supply all the metals and all the miscellaneous inorganic materials used in the world's industries. Collecting and studying minerals as a hobby will usually provide, among other rewards, some understanding of the problems involved in finding and recovering supplies of strategic and industrial minerals.

THE STUDY OF MINERALS

The subject of mineralogy is conveniently considered by subdivision into several categories. In each division, mineral properties are collected and correlated, leading to an arrangement of minerals that simplifies their study.

Early scholars who studied minerals were fascinated by the shapes of mineral crystals, and from this the study of crystallography emerged. Other recognized subdivisions of mineralogy include physical mineralogy, chemical mineralogy, optical mineralogy, determinative mineralogy, and descriptive mineralogy.

Crystallography is the branch of mineralogy concerned with the arrangement of atoms in minerals, and how these arrangements are related to crystal growth, properties, and external forms. Crystallography is a general science with practitioners in many fields, and minerals are but one class of materials studied by crystallographers.

Physical mineralogy deals with the physical properties of minerals. These can be conveniently subdivided as follows:

specific gravity;
mechanical properties, including hardness;
cohesion, including cleavage and fracture;
deformability;
optical properties, including refraction, reflection, luster, and luminescence;
electrical properties, including pyroelectricity and piezoelectricity; and
magnetic properties.

Chemical mineralogy includes consideration of the various chemical properties of minerals, as related to their modes of formation and associations, as well as their occurrence.

Optical mineralogy is concerned with the

Preceding pages: Excellent crystals of sulfur, formed in an area of volcanic activity in Sicily. *Opposite top:* From South Africa, a crystal of diamond, the hardest known mineral. *Opposite bottom:* An emerald crystal—the gem variety of beryl—from Muzo, Colombia.

determination and significance of the optical properties of minerals. This is usually accomplished by microscopic examination, utilizing polarized light. A complete treatment of this complex subject is beyond the scope of this book.

Determinative mineralogy deals with the determination of the nature and composition of minerals by any available means, including physical tests, blowpipe analysis, wet chemical analysis, crystallographic properties, and optical properties. These properties may be arranged in tabular form to facilitate the identification of minerals, after specific properties have been identified.

Descriptive mineralogy brings together and classifies mineral properties, crystal forms, mineral occurrences and associations, applications, and production.

PHYSICAL PROPERTIES OF MINERALS

The usefulness of minerals goes far beyond their obvious importance as sources of metals. In art and industry it is their physical properties, rather than their chemical composition, that makes various minerals desirable and useful. Certain properties give specific minerals value in specific applications.

For example, hardness may be required for abrasive uses, as may certain fracture characteristics; garnet, corundum, and diamond are therefore excellent minerals to use in abrasion. Oil well drillers need a substance with high specific gravity in order to make heavy mud for drilling, and for this application barite is ideal. New applications for minerals, based on their physical properties, are continually being found.

Specific Gravity

Everyone is familiar with the fact that some substances are heavier than others. But to compare any two substances properly, you have to use units of the same size. Imagine, then, a series of identical cubes. The first holds exactly one pound of water at its maximum density, which is attained at a temperature of 4° C. The next cube is made of sulfur, the next, quartz, followed by diamond, galena

(lead sulfide), lead, and gold. If you weighed each of these cubes, you would find that they weigh, respectively, 2.07, 2.65, 3.52, 7.5, 11.37, and 19.33 pounds. Each weight shows the number of times the substance is heavier than an equal volume of water. This is called its specific gravity, which we may define as the *ratio of the weight of a given volume of a substance to the weight of an equal volume of water.* The specific gravity of a substance, abbreviated SG, is of considerable importance in calculating the weight of a given mineral specimen. Remember that a cubic foot of water weighs 62.4 pounds.

It is difficult to measure the volume of mineral specimens exactly, since they often come in uneven shapes. Thus, it is practically impossible directly to compare the weight of an individual specimen with that of an equal volume of water. Fortunately, however, when a body is submerged in a liquid, the force called *buoyancy* acts to reduce the weight of that body. The buoyant force is equal to the weight of liquid displaced, regardless of how irregular the shape of the body is. This means that if a specimen is weighed while suspended in water, it will appear to weigh less than it does in air. The apparent loss in weight is exactly equal to the weight of water displaced by the specimen. This principle provides a convenient and easy means for determining the specific gravity of minerals—provided they are not soluble in water!

Many devices have been developed for determining specific gravity. All of them require an arrangement for weighing a fragment of a mineral first in air (Wa) and then in water (Ww). From these two determinations the specific gravity, SG, can be computed with the following formula: $SG = \frac{Wa}{Wa - Ww}$. In this formula, $Wa - Ww$ is the weight of water displaced by the specimen.

Two instruments for measuring specific gravity have been popular among mineralogists in the past. One of these, called the Jolly balance, uses spring tension to measure weights. All measurements are pointer readings along some kind of graduated scale.

In measuring specific gravity, certain precautions must always be followed. The specimen must consist of pure material, and must

Opposite left: Apophyllite crystal, showing pearly luster on the basal face (New Jersey.) *Opposite right:* Hexagonal calcite crystals displaying stacking habit of tabular plates (Ohio). *Above:* From Quebec, columnar crystals of diopside, with a vitreous luster.

be of a size within the accuracy range of the instrument used. The specimen must be solid, with no sealed air cavities or voids. It must be dry when first weighed, for a sample that is wet gives erroneous readings when weighed in air. It should be handled carefully to avoid breakage, for the loss of even a small fragment between weighings will produce an erroneous measurement. Finally, the sample must be free from adhering air bubbles when weighed in water, since bubbles will add to its apparent buoyancy, resulting in an inaccurate measurement.

The specific gravity of a substance is fairly constant, although it may vary slightly with the chemical composition range of a given mineral species. It is therefore an excellent measurement to make in the process of identifying specimens, especially in the case of gems, because it can be done rapidly and nondestructively.

Hardness

The hardness of a substance is its resistance to abrasion or scratching. There are various ways of measuring hardness, including tests that involve simple scratching. What is actually being measured, indirectly, is the strength of the bonds holding atoms together within the material. A scratch or indentation is a macroscopic result of the separation of material by breaking bonds.

Care must be taken in measuring hardness in order to obtain consistent and reliable results. The surface of a substance to be tested must be smooth, for example. Since a scratch test implies the direct comparison of two materials, the hardness of the known test sample should be accurately appraised.

The simplest and best-known scale for measuring hardness was first proposed by a German mineralogist named Friedrich Mohs (1773–1839). Still widely used in mineralogy, it consists of ten minerals ranked in order of increasing hardness:

The Mohs Scale

1. Talc	6. Orthoclase
2. Gypsum	7. Quartz
3. Calcite	8. Topaz
4. Fluorite	9. Corundum
5. Apatite	10. Diamond

Each mineral in the scale can produce a scratch mark on any mineral with a lower hardness number. Thus, for example, corundum can scratch topaz or quartz, and in turn it can itself be scratched by diamond. Diamond is the hardest substance known, and will scratch any other material on earth.

In making a hardness determination of an unknown mineral, a fresh, unaltered surface should be selected. The corner of one of the Mohs scale minerals is then rubbed across a flat surface of the unknown, and the result examined. What may at first appear to be a scratch may only be some powder from the corner of the Mohs scale mineral, if this is actually softer than the unknown. The powder should be wiped away and the flat surface examined again, preferably with a hand lens, to see if the surface has actually been marked or indented.

Erroneous results will sometimes be obtained if the material to be tested is friable, or powdery. Hard inclusions of other minerals in either the scale or the unknown mineral may also give misleading results.

For hardnesses from 5 to 10, a set of hardness pencils is convenient and effective. These consist of short hexagonal brass rods, with pointed fragments of minerals attached at each end. The scale covers the values of 5, 6, 6.5, 7, 7.5, 8, 9, and 10. When hardness pencils or minerals are not available, the following common materials can be used:

Fingernail: hardness slightly over 2
Copper coin: hardness about 3
Pocketknife blade: hardness slightly over 5
Window glass: hardness about 5.5
Steel file: hardness 6 to 6.5

Another useful hardness-testing device is a scribing tool. The point of this instrument is nearly 6 in hardness, and can be reversed in a small chuck and carried safely in the pocket. The advantage of the scriber over mineral test samples is the small size of the point; the hardness of even a minute grain can be tested and examined with a hand lens.

Our usual concept of a test scale is one in which equal units on the scale vary by equal intervals. This is not true for the Mohs scale. Although the difference between unit intervals for most of the scale is not too large, there is a tremendous break between the last two minerals on the scale, corundum and diamond. The difference in hardness between these two is actually fifty to one hundred times greater than the difference in hardness between corundum and talc, the first mineral on the scale!

The hardness of a mineral varies with the direction of the scratch, because the bond strengths in most crystal structures are not uniform. Usually the difference is not detectable without precision testing apparatus. Some minerals, however, such as kyanite, vary so greatly in hardness that even simple scratch tests are sufficient to detect it. Kyanite may show a hardness of 4.5 in one direction of scratching, while if the scratch is made at right angles to this the measured hardness is 7.

Cohesive Properties

Cleavage

The atoms that make up the structures of crystals are arranged in definite layers or planes, all of which have a fixed angular relationship to each other. It follows, therefore, that if the spacing in any one set of planes is greater than that in any other set of planes, the structure will probably separate more readily along the directions where the spacing is the greatest, or where the cohesive force is at a minimum.

When a mineral can be broken readily along certain crystallographic directions, it is said to have cleavage. For example, if you attempt to break a crystal of halite (rock salt), you will find that it breaks readily, producing three smooth, lustrous surfaces at right angles to each other and in the positions of the cube faces of the crystal. Halite is said to have cubic cleavage.

Cleavages are always designated by the crystal form to which they are parallel. If the *same* cleavage occurs in two or more directions, the quality and appearance of the cleavage surfaces must all be the same. Fluorite, for example, displays four cleavage planes that make an angle of about 70° to each other. These cleavage surfaces define an eight-sided shape called an octahedron, and fluorite is said to have octahedral cleavage. Sphalerite (zinc

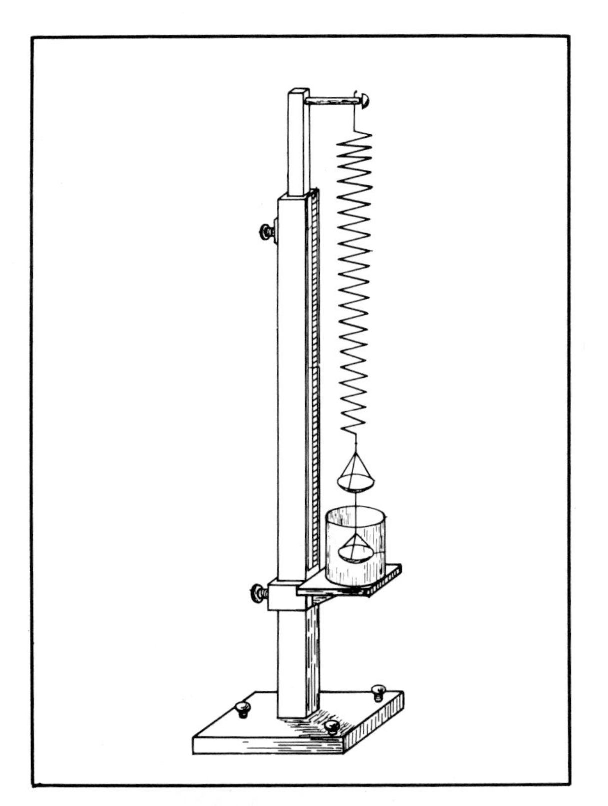

Top: A Jolly balance, for measuring specific gravity. *Above:* Apophyllite crystals, with low specific gravity and hardness of about 5 on the Mohs scale (Centreville, Virginia).

sulfide) has six cleavage surfaces that enclose a twelve-faced solid called a dodecahedron, and sphalerite is said to have dodecahedral cleavage. The mineral gypsum (calcium sulfate), in the variety known as selenite, has three cleavages. Two of these have one direction each and the third has two directions. In this case, each cleavage surface has a luster distinctly different from the others. Calcite, one of the commonest minerals, has rhombohedral cleavage, which may be described as being in three directions not at right angles.

Cleavages are designated according to the perfection of the surface obtained upon breaking, as follows:

eminent cleavage produces perfectly smooth, brilliant surfaces, such as those produced in the micas muscovite and phlogopite;

perfect cleavage yields relatively smooth and lustrous faces, as in the rhombohedral cleavage of calcite;

good cleavage shows surfaces that are well developed and more or less even.

Terms used to express inferior cleavage are *poor, distinct, indistinct,* and *imperfect.* In addition, one may refer to cleavage as being very easily obtained, as in many of the micas, which require very small effort to cleave. On the other hand, orthoclase, one of the feldspars, is not nearly as readily split, and other minerals are still more difficult to cleave.

Parting

Some minerals have an inherent property known as parting, which is believed by many to be due to differential stresses caused by variable dynamic pressures imposed upon a mineral after it has crystallized. The planes along which parting occurs are parallel to some common crystal face, and the parting is named according to the crystal form it parallels. Examples are octahedral parting in magnetite, rhombohedral parting in corundum, and basal parting in diopside.

Fracture

The breaking of minerals along directions other than cleavage or parting planes is called fracturing, and the surface along which the break occurs is called a fracture surface. Such surfaces often have distinctive characters.

When the convex part of the broken surface resembles the shell of certain mollusks, the fracture is said to be conchoidal. The conchoidal configurations may vary in size, from large to small, and in their degree of shell-like appearance. Conchoidal-like surfaces less characteristically developed are called subconchoidal. Such surfaces grade into those said to be either even or uneven. Hackly fractures contain sharp or jagged elevations such as those shown by copper, while a splintery fracture closely resembles the fracture of wood, as shown by spodumene when broken across certain faces. Earthy fracture has the appearance of a break across hard soil and is characteristic of some occurrences of red hematite, bauxite, and granular manganese minerals. Earthy fractures are confined mostly to consolidated, porous, granular mineral aggregates.

Deformability

Minerals can be deformed and broken under stresses that exceed their breaking strength.

Opposite: Top: Rhombohedral dolomite crystals (Spain). *Bottom left:* Pyrite crystal, with notable metallic luster (Island of Elba). *Bottom right:* Limonite, displaying striking iridescence (Georgia). *Above:* Fine color and form in a wulfenite crystal (Arizona).

In refraction, light traveling from air into glass is bent along the path *aobc*. The angle of incidence—the entry angle—is *i;* the angle of refraction is *r*. *P* is the perpendicular.

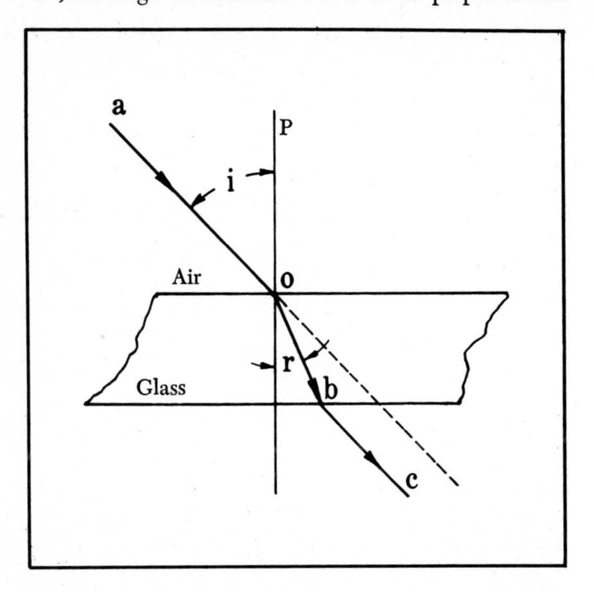

Their reactions to such stresses vary greatly. A number of terms are used to describe the behavior of minerals when subjected to stresses produced under various physical conditions.

Elasticity

Practically all minerals are elastic within certain definite limits of stress. If the mineral has not been stressed beyond the elastic limit, it will return to its precise initial condition when the stressing forces are released. The size of the sample has no influence on the actual elasticity a mineral, or any other substance, possesses. However, the shape of the sample *is* a factor. When subjected to stresses, thick, short pieces can be stressed up to or beyond the elastic limit and can reach the breaking point without any visible bending. On the other hand, a very thin sheet or filament of the same material may be bent or twisted through a large angle before the elastic limit is passed and the breaking stress is reached in the crystal lattice. The breaking point is the stress level when the atomic bonds in the material are not strong enough to maintain its cohesion; thus, breaking occurs.

Quartz is one of the most highly elastic minerals, but it is difficult to obtain in thin sheets or filaments. The minerals that show their elastic properties best are the micas. These can be readily cleaved into very thin sheets that bend easily and return to their original positions when the bending force is released. They are visibly elastic when bent about a line in the plane of the mica plate; but even a hairlike crystal of mica, if elongated perpendicular to this plane, will break quickly under a relatively low stress.

Flexibility

Thin cleavage flakes or slices of some minerals can be bent easily, but even a very small amount of bending will often produce a permanent set—that is, the mineral will not spring back to its original position. Some materials will not break even after being bent 360°. Talc and orpiment are good examples of nonmetallic minerals with a high degree of flexibility.

Ductility

Ductile substances, usually metals, are those that can be made into wire by being drawn through dies. A comparatively few metals occur as native elements, and some of these, such as copper, silver, and gold, are ductile enough to be drawn into wire.

Malleability

Malleable minerals are those capable of being hammered into thin sheets. Gold is the most malleable of all metals and minerals. Sheets approximating $1/200{,}000$ of an inch in thickness —called gold foil—have been produced from gold by hammering it between soft leather.

Sectility

Sectile minerals can be cut with a knife, yielding a shaving. In addition to some of the native metals, a few other minerals can be cut in this manner. The most important of these is argentite.

Brittleness

Brittle minerals are those that can be easily reduced to fragments with a hammer or by crushing under fairly light pressure. Cerussite, for example, is extremely brittle. When the amount of energy or crushing force required to reduce a mineral to fragments becomes very great, the mineral is said to be tough.

OPTICAL PROPERTIES

The physical characteristics of minerals that affect light, and whose effects are perceived only through the eyes, are called optical properties. Only a few of these can be considered here. They are chiefly those properties that are visible and can be determined without the aid of the microscope.

When light passes through a solid substance without being diffused, the solid is said to be transparent. Objects viewed through a transparent material can be seen clearly. Materials that transmit light but diffuse or scatter the light rays passing through are known as translucent. An opaque substance, on the other hand, will not transmit light, even through a very thin edge.

Reflection

One of the most important optical properties of minerals is the reflection of light from the surface of a specimen. If no light were reflected from an object and its surroundings, it would be completely invisible. We all know that some surfaces are much better reflectors than others—silver, for example, or other metal surfaces used in mirrors. Black matte surfaces, on the other hand, reflect little light and are visible chiefly in relation to the materials around them that do reflect light. When we look at a fine mineral specimen, we see it in terms of the amount and kind of light reflected differentially from its surfaces that face the observer.

The phenomenon of reflection follows very definite and well-known laws. Specifically, *the angle of incidence equals the angle of reflection.* The incident angle is the angle an incoming light ray makes with an imagined line *perpendicular* to the reflecting surface. The reflected angle is the angle the outgoing, reflected ray makes with this perpendicular. These two angles are always equal.

The large number of minute surfaces on tarnishes or oxide coatings on some mineral specimens scatter the light in so many directions that good reflections are unobtainable. Only occasionally are the surfaces of the larger crystals of minerals sufficiently plane and smooth to act as small mirrors. The fine powders of most nonmetallic minerals are white, due to the scattering of light, by reflection, from the surfaces of the very great number of fine particles.

Refraction

When a beam or ray of light falls upon an object that is not opaque, part of the light is reflected, as described above. If the mineral is transparent, a large part of the light enters and passes through the mineral. The path followed through the substance is a straight line, but the beam is bent or broken at the points where it enters and leaves the mineral. This is the same phenomenon that makes a stick appear to be broken when it enters a pool of clear water. It is known as refraction.

Refraction occurs only when light passes between substances of different optical densities, such as from air into glass, or vice versa. The velocity of light changes as it passes between these substances. The ratio of the two velocities—the velocity in air and the velocity in another substance, such as glass or a transparent mineral—is constant and provides us with the *index of refraction,* a measurement of the degree to which a substance can bend a beam of light. This is an important property and is constant for any homogeneous transparent substance. It is most useful in the identification of minerals.

Total Reflection

Another phenomenon of light that is important in mineralogy, and especially so in gemology, is that of total reflection.

Light, in passing from one medium to another of lower refractive index, may be reflected rather than refracted. The refraction angle at which this occurs is called the *critical angle.* This is important in gemology because the measurement of the refractive index of gems is based on the measurement of critical angles.

Critical angle also affects gem brilliance. Ideally, light entering a cut gemstone should be reflected back internally, and only be allowed to exit through the top of the stone. This can be managed by placing facets at proper angles to take advantage of the stone's refractive index.

Luster

When a beam of light falls upon a mineral, some of the light is absorbed; if the mineral is transparent, a large part of the light passes on through the substance, but there is always a portion of light reflected from its surface. Some opaque minerals, such as pyrite, have mirror-like surfaces and reflect most of the incident light. Depending upon the quantity of light reflected from the surface of a mineral, it has either a bright or dull appearance, which is referred to as its luster. Opaque minerals, especially the metallic ones with smooth, clean surfaces, reflect most of the light. They appear extremely bright and are said to have a *metallic* luster. The surfaces of many of the metallic minerals tarnish rather readily and become dull and lack the bright reflections. Their luster is said to be a *dull metallic* one. If the mineral is nonmetallic but is opaque, the luster is described as being *submetallic*. A *nonmetallic* luster is associated with the transparent minerals or those that are transparent only in very thin edges, even though they may be deeply colored.

The highest luster of the nonmetallic minerals is close to that of the diamond and belongs to those minerals having an index of refraction above about 2.00. They are said to have an *adamantine* luster. Between an index of refraction of 1.75 and 2.00, the luster is generally of a *subadamantine* appearance. Glass is said to have a glassy or *vitreous* luster. Minerals having an index between 1.45 and 1.75 usually have a vitreous luster, while those with an index below 1.45 usually have a *greasy* luster, like the appearance of solidified fats and oils.

Lusters are sometimes named for their similarity to some well-known objects. Among these we have the *silky* or *satiny* lusters resembling silk or satin—a striking example is that of a variety of gypsum known as satin spar. Some minerals possess a *pearly* luster, resembling the mother-of-pearl of mollusks. The term *resinous* luster has been applied to the fossil resins such as amber, and sometimes to minerals with a brownish-yellowish appearance, such as sphalerite, which has in reality an adamantine luster.

Determination of the luster of a mineral must be arrived at in relation to a fresh surface, such as a smooth cleavage or fracture surface of an unaltered specimen. Polished surfaces or bright crystal faces are also good spots for determining the luster. Luster is not an invariable determinative property, as there are no sharp lines dividing the various kinds of lusters.

Color

Color is impossible without light, and visible light consists of a small section of electromag-

Above, top right, and *right:* Excellent crystals of chalcanthite, sulfur (Sicily), and garnet (Roxbury, Connecticut). *Opposite:* Fluorescing minerals photographed under ultraviolet light to show their characteristic intense colors. The green mineral is willemite, and the red, calcite, both from Franklin, New Jersey; the yellow is wernerite, from Quebec.

netic vibrations of comparatively short wavelengths. When all wavelengths are received simultaneously by the eye in the proportion in which they are emitted by the sun, the light is said to be white and of daylight quality. From this it is obvious that the usual color appearance of an object is due to the absorption of certain wavelengths of the incident white light and the reflection of the remaining wavelengths, which combine before reaching the eye to produce the particular color sensation associated with any given object.

If you look at two specimens of the same mineral in your own cabinet or in the mineral hall of a large museum, you may be surprised to find considerable variation in their color. The hue, tone, and intensity of color frequently vary unless the specimens come from the same mineral deposit, and even then there are often wide variations in the same species or even in the same specimen. There is, however, a close similarity in the color of specimens of some minerals, which makes it an important physical property, when used with caution, in the identification of minerals.

Dispersion

Dispersion is the ability of a material to separate white light into component wavelengths, causing it to appear as a band of colored light. The greater the difference in refraction between red and violet—the opposite ends of the spectrum—the greater will be the "spread" of colors emerging from the material.

Double Refraction

When a clear, transparent piece of a mineral such as calcite is placed on a printed page, a double image of the printing will appear. In this phenomenon, called double refraction, the beam of light entering the material is split into two beams, which take different paths through the mineral—hence the double image. The particular crystal structure of any mineral exhibiting double refraction breaks up the light in a particular way. The amount of splitting and the directions of the two beams depend on the nature of the crystal. The splitting is a function of the direction the light takes through the crystal and the structure of the crystal itself.

Only amorphous substances and those crystallizing in the isometric crystal system do not produce double refraction. All other crystals do produce the phenomenon, to varying degrees.

Light travels in waves, like ripples on water. Such ripples extend at right angles to

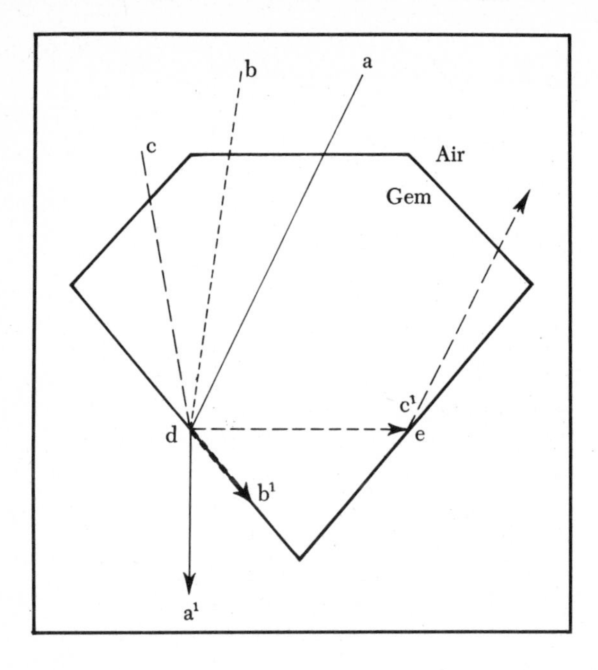

their direction of travel. But the "up and down" motion of such waves is like a vibration in only one direction. Light generally vibrates in *all* directions at right angles to its direction of travel. In double refraction, the vibration direction of the two beams of light are at right angles to each other. However, light that is *polarized* has been forced to vibrate in only one plane. When such light passes through a transparent mineral, it reveals colors that are typical for that mineral. The polarizing microscope thus is used by mineralogists to identify mineral specimens.

Pleochroism

We have seen that some minerals break up a beam of light into two parts. Each of the rays travels at a different velocity, and the various wavelengths of light in the beams are treated differently along the paths of the doubly refracted beams. In some minerals, such as the more deeply colored tourmalines, the ray that vibrates in one direction is almost completely absorbed. In other minerals, only certain wavelengths are absorbed from either one or both rays. Hence, some minerals will have a variation of color when viewed along the vibration directions of the two refracted rays. Minerals that affect light in this manner are said to be pleochroic.

Streak

The streak of a mineral is the color of its powder, and is so named because the powder is usually obtained by drawing a corner of the mineral across a piece of unglazed porcelain plate, called a streak plate. The plate has a hardness close to 7, and in this manner a streak of powder is obtained on the plate for minerals with a hardness less than that of the plate. The colored powders are easily seen on the white surface. Some very fine-grained sandstones, such as novaculite, make excellent streak plates.

Powders of the minerals harder than feldspar—the source of clay used in porcelain—may be obtained by finely crushing a piece of the mineral on a steel anvil and then moving some of the finely ground material to a piece of white paper for observation. All white minerals and most of those with a light to medium color produce white streaks. All metallic minerals produce colored, black, or metallic streaks; the others usually produce a grayish streak. Although the color of a mineral may vary between wide limits, the color of the streak is surprisingly constant and characteristic of that mineral.

Tarnish

When the bright surfaces of minerals are dulled by an oxide film or other coating, they are said to be tarnished. Such coatings often show distinct colors, which are usually different from the color of the mineral. Some of the copper minerals, such as bornite, readily develop a tarnish on freshly broken surfaces of a specimen when those surfaces are exposed to the air.

Iridescence

The many-colored effects that often appear on the surfaces of objects—usually caused by the interference of light rays in the outer surface layers of very thin films or extremely narrow fractures—are often referred to as iridescence. Such color variations are frequently shown by oil films and mother-of-pearl. Iridescence can appear either on the surface, from coatings and films, or from within the body of a transparent material. Tarnished surfaces are often iridescent.

Light may be totally reflected when moving from one medium to another of lower refractive index. Path *aa'* is normal refraction; *bb'* is the grazing incidence. Along *cc'*, the path of total reflection, the light is reflected back into the material. In a cut gem, as here, the light may be reflected twice, at points *d* and *e*, before it emerges.

In dispersion, light passing through a solid is separated, emerging as a band of colored light at *ab*. Each color has its own refractive index for any given substance: here r_r for red and r_v for violet.

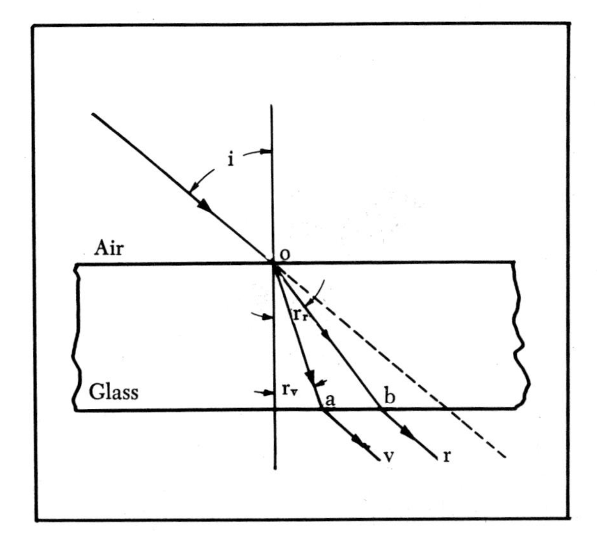

Opalescence

Some minerals display a whitish to faintly colored pearly iridescence from the interior of the mineral. This phenomenon, called opalescence, is usually caused by a large number of minute inclusions or by a variation in the crystal or other structure. The name is derived from the common opal, which frequently shows the effect exceptionally well. It is also shown by moonstones and some types of corundum.

Play of Colors

The play of colors exhibited by minerals when rotated in a beam of white light is caused by various phenomena—dispersion, iridescence, opalescence, interference, total reflection, and others—which cause the white light to separate into its component colors. Moving the specimen causes a rapid change of the colored light rays coming from the specimen. Cut stones, such as diamond, zircon, colorless rutile, and precious opal, exhibit a remarkable play of colors.

Change of Colors

In contrast to the quick change in color of diamond and opal, some minerals, upon being rotated in white light, change their color gradually. Among these are the well-known crocidolite, from Cape Province, South Africa, (commonly called tiger's-eye), and the cymophane variety of chrysoberyl, the cat's-eye, when cut *en cabochon* and polished. These and other minerals possess the property of chatoyancy—that is, they have a changeable appearance or color.

Asterism

A few minerals, especially those that crystallize in the hexagonal or monoclinic systems, produce under certain circumstances a well-developed stellate figure of reflected or transmitted light. In some instances, as with corundum, the phenomenon appears in reflected light, while in other minerals, such as phlogopite, a stellate arrangement appears in the transmitted light. The effect is usually due to the reflection of light from minute capillary inclusions oriented along prominent crystallographic directions. The stellate rays are at right angles to the inclusions; the more intense rays are at right angles to the direction of the most prominent inclusions. The number of rays depends upon the number of sets of directional inclusions. The three directions in corundum and quartz, for instance, give rise to six-rayed stars. The gemstones of these minerals are called star ruby, star sapphire, and star quartz. In cymophane and some other minerals there is but a single set of inclusions, which gives rise to a single bright line. Minerals of this nature produce the so-called eye stones. Some phlogopite from Burgess Township, Ontario, may produce stars with as many as thirty-six well-defined rays.

Luminescence

Under certain conditions, some objects emit light below the temperature of incandescence. This property is known as luminescence, which may be produced by friction (triboluminescence), by a slight rise in temperature (thermoluminescence), and by shortwave electromagnetic irradiation (fluorescence). Of these the last is the best known.

Fluorescence

The property of fluorescence was first described by Sir George Stokes in 1852, who named it after the mineral fluorite, which best displayed it. The phenomenon remained chiefly a laboratory demonstration until early in this century, when inexpensive lamps, radiating a relatively intense ultraviolet light, were developed. Such lamps now cover a wide range of

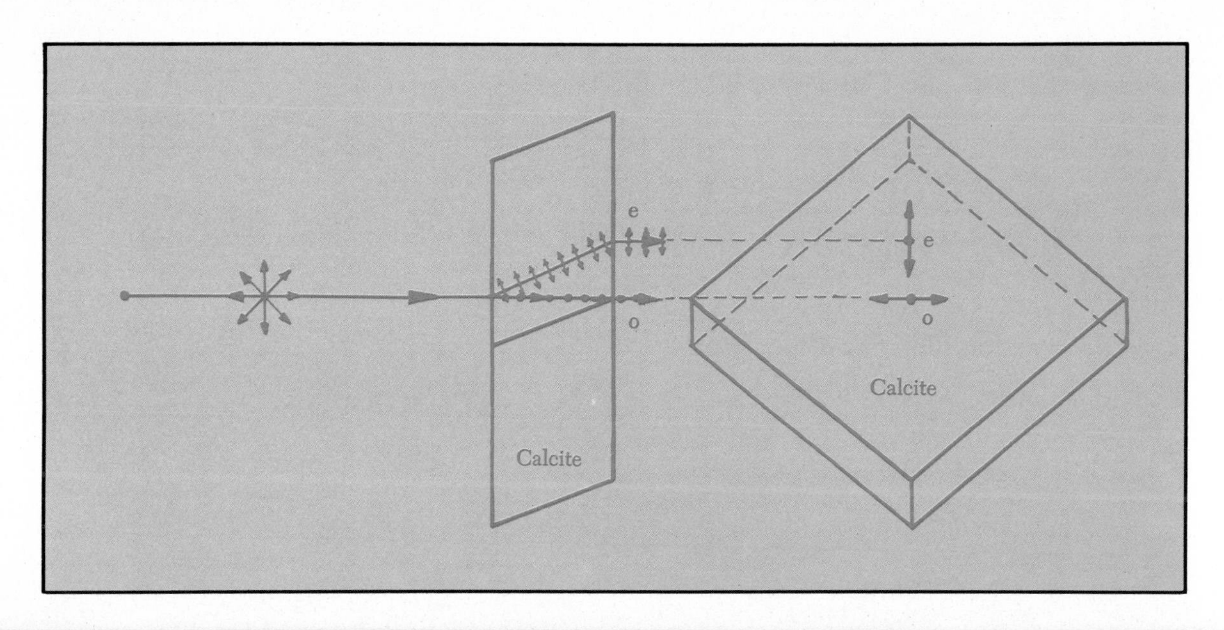
e
e
o
o
Calcite
Calcite

radiations in the ultraviolet region and have brought the phenomenon of fluorescence to most mineral cabinets. Not all specimens of a given mineral species will fluoresce—to do so, the particular specimen must possess some special property. It is thought that the presence of manganese, which is proxying in the crystal lattice for some of the essential atoms, may be one activating material. The hue of the visible fluorescence of a mineral is usually confined to a relatively narrow range of the spectrum, but the tone and vividness of the colors vary greatly, depending on the wavelength and intensity of the ultraviolet light. To first stand in a showroom of fluorescent minerals and alternately turn on and off the tungsten and ultraviolet lights is an experience that the novice will long remember; the change of lights produces a striking color change, from the usual rather drab appearance of the minerals in white light to the amazingly vivid and intense fluorescent colors in shortwave light. Among the many common minerals that are often fluorescent under most of the ultraviolet lights are willemite, wernerite, autunite, scheelite, fluorite, and calcite from Franklin, New Jersey.

The phenomenon of fluorescence has been incorporated into modern lighting used in industrial buildings, stores, and large offices, as well as at home. The television picture screen depends entirely upon fluorescent chemicals.

Phosphorescence

Some minerals continue to emit visible light after the exciting rays have been turned off. This property—actually a continued fluorescence—was first discovered in connection with phosphorus, which glowed in the dark after being exposed to the ultraviolet light of daylight. For this reason a mineral that continues to emit light or fluoresce after the exciting rays are turned off may rightly be said to be phosphorescent.

Triboluminescence

A few minerals from certain localities emit light that appears in a sparklike manner at the points along which friction occurs, either by hammering, crushing, rubbing, or scratching. These minerals include sphalerite and feldspars and the property is called triboluminescence.

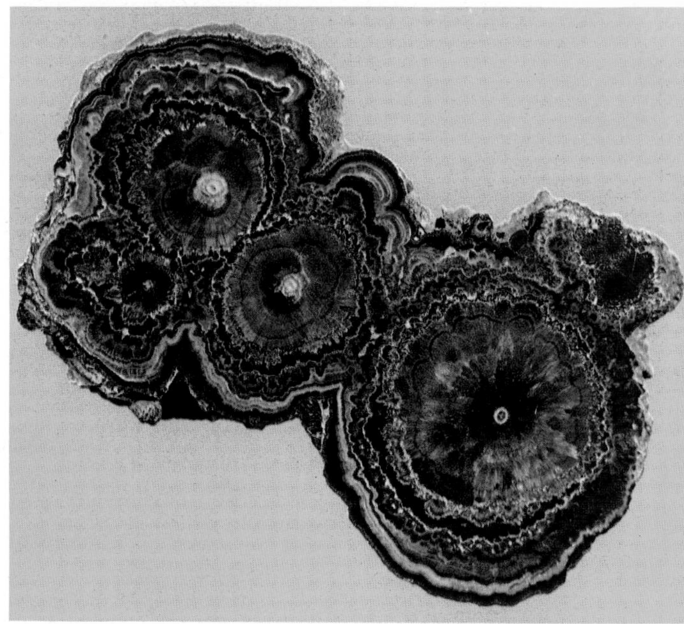

Opposite top: Light normally vibrates in all directions perpendicular to its line of travel. Minerals that are doubly refracting—such as calcite—divide light into two beams. One beam continues in the same line of travel, while the other is refracted. Each beam then vibrates in only one plane, at right angles to the other beam, which results in the appearance of a double image. *Opposite bottom:* Crystals of calcite, a mineral that takes many different forms. *Top:* Gold, a soft mineral with high specific gravity, crystallizes in the isometric system—here in an octahedron. *Above:* Fine color in a piece of azurite with concentric banding.

Thermoluminescence

When a mineral becomes luminescent as it passes through a range of temperatures below red heat, it displays the property of thermoluminescence. The temperature ranges are usually below 200° C. Fluorite from some localities shows the property in an excellent manner in green and lavender colors. The fluorite that emits a green color upon being heated has been called chlorophane. You may readily observe the property by heating fragments in a dark room on the bottom of an inverted electric iron or a hot plate with enclosed heating element.

PROPERTIES THAT AFFECT THE SENSES

In addition to optical properties, which can be perceived with the eyes, minerals display characteristics that affect the other senses. These include odor, taste, and feel. However, such characteristics are difficult to quantify, and so are of limited usefulness in mineral identification. In only a very few cases are they sufficiently diagnostic to be useful in a practical sense.

Odor

The volatility of minerals in general is so extremely low that the presence of vapors released at ordinary temperatures, even in the proximity of large amounts of the finely powdered material, cannot usually be detected by the sense of smell. With few exceptions, minerals are odorless. When an odor can be detected in minerals or rocks, it is often due to the presence of some volatile foreign substance, usually of organic origin, that has permeated it to a greater or lesser degree. An example is hydrogen sulfide, which produces a fetid odor in some minerals and rocks. In a like manner, porous rocks and the enclosed minerals may contain asphaltic or bituminous products, the release of which produces a bituminous odor.

When sufficient heat is applied to some minerals by rubbing or hammering, small amounts of the minerals may be vaporized and produce an odor similar to that of garlic. Selenium minerals when heated produce an odor resembling that of horseradish. Clay if breathed upon, heated, or sprayed with hot water emits a distinctive odor, called argillaceous, meaning claylike.

Taste

When placed on the tongue, a small piece of the more or less water-soluble minerals produces a rather characteristic taste, which is useful in identification. Halite is said to be salty; if an acid is present, it produces a sour taste. Alums and sulfates of metals, such as iron melanterite, produce an astringent taste; nitrates of potassium and sodium produce a cooling taste; soda and potash give an alkaline or sweet taste; and the taste of epsomite is bitter.

Feel

Certain very compact varieties of serpentine and sepiolite (meerschaum), as well as talc and soapstone, possess a distinctly smooth feel, which is often quite diagnostic. Others, such as sandstone, chalk, and diatomaceous earth, have a rough or raspy feel, which varies from the coarse- to the fine-grained phases.

ELECTRICAL AND MAGNETIC PROPERTIES

Magnetism

The magnetic property of lodestone is frequently quite noticeable in the field. The powder produced by striking the mineral with a hammer consists of crushed pieces that will often "stand up" around the bruised spot on the mineral. Lodestone, a variety of magnetite, is polarized magnetically and acts in the same manner as an ordinary steel magnet.

Pyroelectricity and Piezoelectricity

Pyroelectricity is the development of an electric charge on opposite ends of a crystal when the crystal is heated. A similar effect, induced by mechanically stressing a crystal, such as by intense pressure, is called piezoelectricity. The phenomena can only be produced in crystals that lack a center of symmetry.

Some minerals are strongly piezoelectric. Quartz, because of its remarkable piezoelectrical character, is of great importance in the production of sensitive pressure gauges—especially for controlling the transmitted wavelengths of radio and television broadcasting stations.

COMPOSITION OF THE EARTH'S CRUST

The composition of the earth's crust has been under systematic investigation for more than one hundred years. Before accurate methods were developed for the chemical analysis of rocks, opinions concerning their composition were hardly more than speculative. During the latter part of the nineteenth century and the early decades of the present century, many chemical analyses of rocks and their statistical evaluation were published. This work continues to the present day. Exhaustive evaluation of these analyses and the elimination of poor data have led to a steadily improving picture of the earth's chemistry. Compilations of rock data have given us a much clearer insight into the average composition of the earth's crust, down to a depth of about ten miles.

Most of the chemical elements are quite scarce, present in the crust in amounts below $\frac{1}{100}$ of 1 percent. Those elements present in greater amounts are listed below:

Element	Percentage	Element	Percentage
Oxygen	49.52	Manganese	.08
Silicon	25.75	Sulfur	.048
Aluminum	7.51	Barium	.047
Iron	4.70	Chromium	.033
Calcium	3.39	Nitrogen	.030
Sodium	2.64	Fluorine	.027
Potassium	2.40	Zirconium	.023
Magnesium	1.94	Nickel	.018
Hydrogen	.88	Strontium	.017
Titanium	.58	Vanadium	.016
Chlorine	.188	Cerium, yttrium	.014
Phosphorus	.12	Copper	.010
Carbon	.087	All others	.032

It is surprising to note that, of the ninety-two chemical elements occurring in the earth's crust, only eight constitute more than 1 percent each, by weight, of the crust, and all of them are important in our economy. Even a small, basic mineral collection is likely to contain minerals made up of the more abundant elements. An unusual and unique collection would consist of minerals representing every chemical element found in the earth's crust—but it would be difficult to collect!

In double refraction, the image appears to be doubled when it is placed under a rhombohedral piece of calcite crystal.

Looking at the average composition of igneous rocks, we find that relatively few minerals are included. These are quartz (12 percent), feldspars (59.5 percent), pyroxenes and amphiboles (16.8 percent), micas (3.8 percent), and accessory minerals (7.9 percent). The average mineral collector, on a trip to almost any mineral locality, can easily find specimens of most or all of the above minerals, in addition to many other mineral species.

The collector is immediately confronted with the fact that only a few of the nearly 2,500 known species can be easily found in the field. Most mineral collections, even large ones, are limited in the variety they can contain. Collections with more than a thousand species are limited mostly to large universities and museums. The mineral collector must therefore take steps to increase his chances of finding interesting, less common mineral species, such as by traveling to specific localities noted for their mineralogical abundance. Suggestions for field trips that should yield interesting specimens are offered in Chapter 6.

2. Lattices of Beauty

Minerals are scattered throughout the earth's crust. When a sufficient number of chemical elements are concentrated because of various geological processes, the atomic elements present may form solids that we call crystals. Study of these mineral solids is the science known as crystallography.

Because these crystals are often brilliantly beautiful, they have been called the flowers of the mineral kingdom. For centuries, the student of mineralogy has been fascinated by the crystals he has found in the earth's rocks. The word "crystal" itself is from the Greek *krystallos*, meaning ice. It was the word the Greeks used to describe the splendid groups of quartz crystals they found in the Alps—for these early scientists believed that quartz was water frozen so solid it could never thaw.

Mineral crystals come in a great variety of sizes. Some of the prized specimens on exhibit in museums are less than an inch high. Some are egg-size. Some are considerably larger. Students know that an examination of the surface appearance of a crystal—its outer form, its color, and certain other visible characteristics—sometimes can lead to an identification of the mineral species to which it belongs. More often, a study of its internal structure is required. This study of crystals has appeal not only to the lover of beauty, but to the scientist, the prospector, and the collector.

To understand the structure of a crystal, it is necessary to recognize the geometric elements involved and to appreciate the atomic arrangements by which the crystal developed and grew.

CRYSTALS

A crystal is a solid bounded by natural plane surfaces, or faces, which are developed at the time the solid substance is formed. The crystal form outwardly expresses the definite internal atomic arrangement of its constituent atoms, which are acquired during growth.

A crystal face is a natural surface, which is usually smooth, plane, and reflective, unless it has been blemished by the corrosive action of solutions during or after growth, or marred by contact with adjacent crystals. Plane surfaces that are cut and polished artificially on gems are known as facets. When mineral specimens are polished, their value is actually most frequently reduced, rather than enhanced—for the value of a fine crystal specimen is in the uniqueness and beauty of its natural perfection. This is adversely affected by artificially created surfaces, even though rough and irregular material may be removed.

Formation of Crystals

When a substance passes from a gaseous or liquid state to that of a solid, the result is normally a crystal. Solids may form directly from gases or liquids of the same chemical composition, as in the case of steam, which condenses to water and then freezes to solid ice. Crystals may also form when material dissolved in a fluid precipitates because of a change in temperature or other variable. An example of this is the evaporation of seawater, causing the formation of crystals of salt.

The growth of some crystals can even be seen with the naked eye. Ice crystals can be seen forming on windowpanes or sidewalks, for example. Snowflakes are formed by the crystallization of ice directly from water vapor in the atmosphere. Most minerals will form crystals when circumstances are favorable.

The forces that cause crystals to form are similar to those present within the crystals themselves. Consider, for example, water. We are familiar with three different forms of this simple substance: vapor (steam), liquid, and solid (ice). Each form is stable, and tends to form spontaneously in a specific range of temperature and pressure conditions. In water vapor, the molecules have a great deal of energy and tend to fly about, colliding with each other and filling any space into which they are introduced. This condition is typical of high temperatures, above 212° F. Below 212°, however, the electrical forces of attraction between water molecules become stronger than the vibrational energies that keep vapor molecules in constant motion. Liquid water therefore forms, in which the molecules cannot fly about randomly, but are still free to slip and slide over one another. This limited freedom of motion gives liquid water the properties that make it so important to life on earth. Liquid water is the most stable form until the temperature reaches 32° F. Below this tem-

Preceding pages: Vanadinite crystals (hexagonal system) from Arizona. *Opposite: Top:* Orthorhombic crystals: *left,* twinned aragonite (Scicily); *right,* cuprite (Australia). *Bottom left:* Chabazite (hexagonal system), with stilbite and calcite (Nova Scotia). *Bottom right:* Fluorite cubes (isometric system), from Elizabethtown, Illinois.

perature, the vibrational energy of the water molecules is smaller than the attractive forces between the molecules, and the movement of the molecules is finally restricted completely. Thus, ice forms, in which the water molecules have relatively fixed locations to one another. They can still vibrate somewhat, but this motion is severely limited.

The formation of a solid, or crystal, is characterized by the loss of heat or vibrational energy, and the settling of atoms and molecules into a fixed, geometrical arrangement. Such a pattern is the most characteristic feature of solids, and so all solids are, by definition, crystalline. Substances such as glass are *rigid,* but not solid, because they have no fixed, internal arrangement. A glass could be considered a supercooled liquid, rather than a solid.

CRYSTAL STRUCTURES

In every crystal there is a fixed, definite, and characteristic arrangement of the constituent atoms. The external shape or form of the crystal is determined by the nature of the atoms and the way they are arranged, as well as by the particular physical and chemical conditions in which the crystal is formed. The pattern, or geometric grouping of the atoms, is known as the crystal structure of the material. All crystal structures are three-dimensional arrangements of atoms.

Electrical attractive forces hold atoms in a crystal together. Different atoms have different associated forces, and in a crystal the forces of attraction tend to vary with direction. This variation accounts for the directional properties of crystals, such as cleavage, hardness, electrical properties, and optical properties.

DESCRIPTIONS OF CRYSTALS

The formation of crystals and the laws governing crystal structures are extremely complex. In past centuries there was no way directly to observe the atomic structures of crystals.

It was originally theorized that atoms did indeed exist. Moreover, if they did, they would have to be arranged in some kind of three-dimensional patterns to account for observed crystal shapes and properties. If you consider a crystal structure as an abstract entity, and mentally replace each cluster of atoms in the structure with a single point, you have a three-dimensional dot pattern that represents the structural arrangement. Such a pattern is called a *lattice,* or *space lattice.*

There are possible only fourteen types of lattices, and although they are all unique in their properties, several lattices have the same basic shape. It turns out that six basic shapes may be accounted for. These give rise to what are known as the *six crystal systems.*

Space is three-dimensional, and any object within it can be described by reference to three *axes,* or reference lines. For example, if you are sitting in a room, and choose one corner of the room as a starting point, you could describe your location as so many feet from the corner along one wall, then so many feet into the room at right angles from that wall. If somehow you could float up into the middle of the room, you would need a third measurement to indicate your height above the floor.

In crystals, any point on the lattice can be considered a starting point, and the dots forming the lattice could be connected by lines.

These lines would reveal the lattice as a series of identically shaped boxes.

Remember that the box does not have to be square or rectangular, with the lattice lines meeting at right angles. Other angles are also allowed. But all the descriptions are tied to the assumption of a lattice point as a starting point, and labeling everything with reference to three imaginary *axes of reference,* or *crystallographic axes.* Any crystal can also be described in terms of these axes.

The differences between the crystal systems are usually expressed by the variation in the number, length, and relationship of the crystallographic axes to each other.

Crystal Systems

Crystals in the *isometric system* have three axes of equal length, oriented at right angles to each other. The distances between atoms vary for different minerals, but spacings along the crystal axes are uniform.

In the isometric system there are always

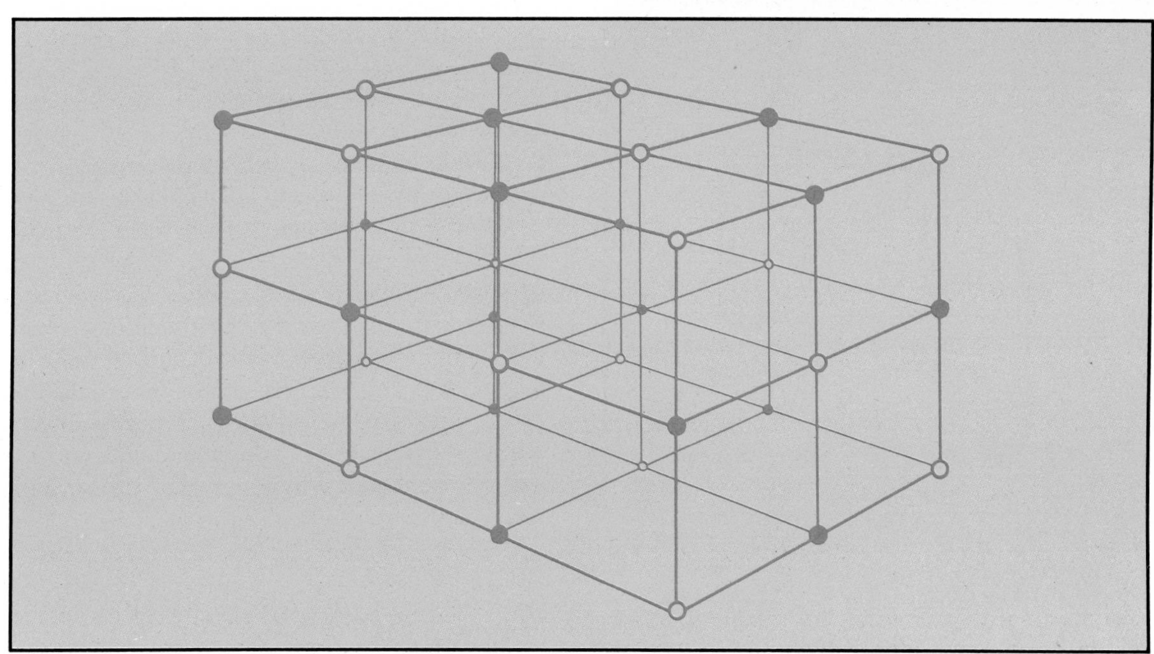

Crystals in the isometric system. *Opposite:* Galena (Joplin, Missouri). *Top:* Copper (Michigan). *Above:* A space lattice, the structure of a crystal. Black and white circles represent different kinds of atoms. Shown here is an isometric crystal structure, such as halite—rock salt—which has two kinds of atoms, sodium and chlorine. Other minerals may contain many more kinds of atoms.

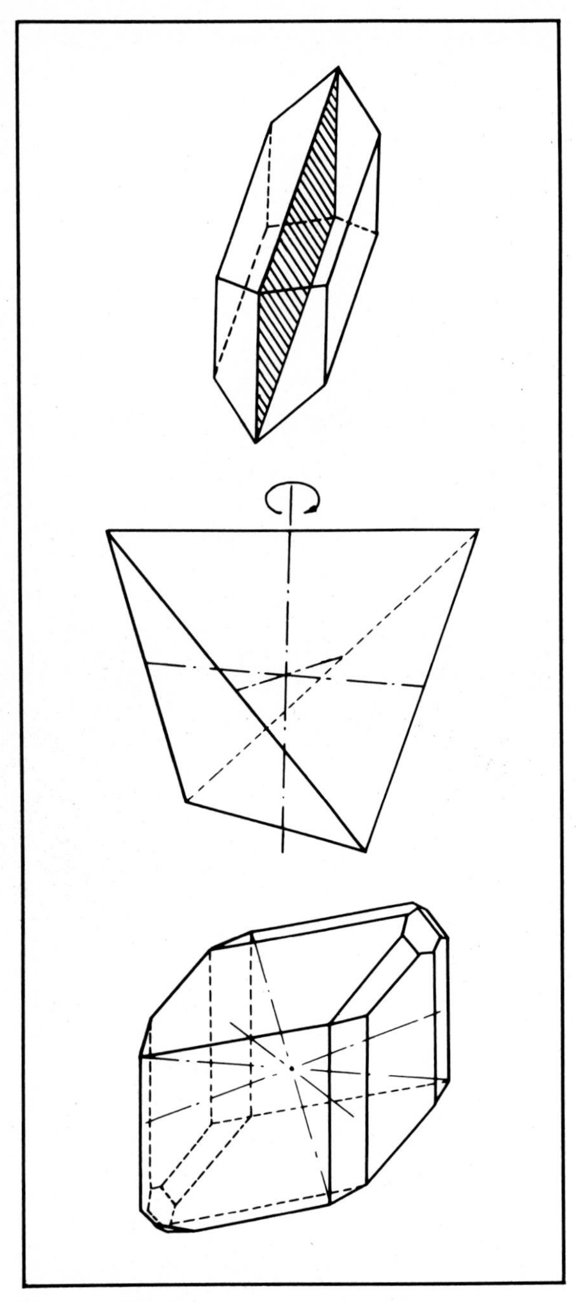

Elements of crystal symmetry. *Top:* Plane of symmetry in a monoclinic crystal; opposite halves of the crystal are mirror images. *Center:* Three axes of symmetry (dot-and-dash lines) in an isometric crystal. *Bottom:* Center of symmetry, with features repeated on opposite sides of the crystal, on a line through the center. *Opposite left:* Tetragonal crystal of chalcopyrite on a quartz matrix (Ellenville, New York). *Opposite right:* Rare twinned crystals of silver, in the isometric system (Kongsberg, Norway).

three or more lines through the crystals that are alike in all respects; that is, they terminate in like faces, edges, or angles. There are no unique lines in any crystal form or combination of forms in the isometric system. Some minerals that crystallize in this system are garnet, pyrite, fluorite, sphalerite, and boracite.

In the *tetragonal system,* there are two equal axes—standardly labeled a_1 and a_2—intersecting at right angles in a plane, with a third axis—labeled *c*—that is longer or shorter and perpendicular to the plane of the *a* axes. Minerals crystallizing in this system include vesuvianite, cassiterite, zircon, wulfenite, and chalcopyrite.

In the *hexagonal system,* there are three axes of equal length lying in a plane, intersecting at angles of 60°, with a fourth axis either longer or shorter at right angles to this plane. The ratio of the length of the fourth axis to that of the three equal axes is always characteristic for any mineral species in this system; its value is constant for any given mineral that crystallizes in this system. The hexagonal system may be divided into a hexagonal and a trigonal division, depending on whether the vertical axis is one of six-fold or of three-fold symmetry. Examples of minerals in this system are beryl, zincite, apatite, pyromorphite, calcite, tourmaline, quartz, and dioptase.

In the *orthorhombic system,* there are three unequal axes, all at right angles to each other. The third—labeled *c*—is usually oriented vertically, with the second axis—*b*—running from left to right. Because the axes are of unequal length, the symmetry is greatly reduced. The planes and axes of symmetry are unique and noninterchangeable. Typical minerals in this system are topaz, barite, olivine, sulfur, stibnite, hemimorphite, and epsomite.

The *monoclinic system* has three axes of unequal length. The vertical axis is called *c,* and the front-back axis, usually designated *a,* is inclined to the *c* axis. The horizontal axis, *b,* is perpendicular to the plane of the *a* and *c* axes. The acute angle of inclination between *a* and *c* is designated beta (β). The characteristic elements for this system are the axial ratios $a{:}b{:}c$ and the angle beta. The maximum symmetry in the monoclinic system is a two-fold axis coinciding with *b,* a center of symmetry,

and a plane of symmetry perpendicular to *b*. The *prismatic class* of this system is well represented among the mineral species. Examples are gypsum, the rock-forming mineral orthoclase, diopside, muscovite, datolite, augite, and realgar.

In the *triclinic system,* the crystal faces are referenced to three axes of unequal length, all of which are inclined to each other. The characteristic elements include the three axial lengths and all three interaxial angles. There can be only two classes of symmetry in this system; in one, the maximum possible symmetry is limited to a center, while the other class is without symmetry. Examples of minerals in this system are albite, rhodonite, and kyanite.

CRYSTAL SYMMETRY

The regularity of shape and arrangement of the faces on good crystals soon becomes obvious to the person beginning the study of minerals. It is also striking that the opposite halves of most crystals are alike. In other words, the crystal is symmetrical with respect to a plane, or planes, passing through the ideal center of the crystal. Such imaginary planes are called *planes of symmetry.* They are always parallel to some prominent crystal face of the class to which the crystal belongs. *Crystal classes* are derived from the ways objects can be arranged in space about a point. They arise from a consideration of different types of symmetry.

In addition to the various kinds of symmetry planes, crystals usually have one or more *axes of symmetry,* and sometimes a *center of symmetry.* An axis of symmetry symbolizes repetition by rotation. If you lay out a square on the floor, put an identical red marble at each corner, and stand in the center of the square, you can visualize a four-fold rotation axis. As you turn about in the square, each time you face a corner you see an identical red marble. You see such a marble four times in a complete rotation. Your body would then correspond to the location of a four-fold rotation axis, and, indeed, the properties of a square indicate the presence of a four-fold rotation axis at its center, oriented perpendicular to the square.

For geometrical reasons, only a few types of symmetry can exist in crystals: two-fold, three-fold, four-fold, and six-fold. A one-fold axis exists in every crystal, even if no other symmetry is present.

A *center of symmetry* causes repetition at opposite ends of a line drawn through the center, and at equal distances from the center. Imagine a cubic box with a point at the exact center. The diagonally opposite corners of the box would be related by a center of symmetry. The same relationship would exist for diagonally opposite edges, faces, or any other points on the surface of the box.

Crystals may contain one or more of these elements of symmetry—planes, axes, center—but there are physical restrictions on the number and types of such elements that can coexist *simultaneously* in a given crystal. In fact, there are only thirty-two combinations of symmetry elements that are possible in real crystals. These give rise to the *thirty-two crystal classes.*

CRYSTAL FORMS AND FACES

Crystal faces can be described by referring to the crystallographic axes. Each crystal class is characterized by the possible presence of specific types of crystal forms. The term *form,* to a crystallographer, means *the set of all the faces that have the same cutting distance, or intercepts, along the axes, and that occupy all such positions relative to the crystal axes as required by the symmetry of the crystal class to which the crystal belongs.*

For example, consider the cube. Passing through the center of opposite cube faces runs a four-fold symmetry axis. If a crystal face exists parallel to this axis, the presence of the axis means that this face must be repeated four times around the crystal. In other words, if the cube is a box, the axis runs through the top and bottom and the symmetry dictates that the cubic box must have four sides.

A form is thus a *set* of faces, all of which

Above: From the top, left to right: Garnet (Maine), barite (Cumberland, England), diopside (Quebec); galena (Cumberland, England); calcite (Taxo, Mexico), malachite (Bisbee, Arizona), sand-calcite crystals.
Opposite: Isometric crystal of halite from Carlsbad, New Mexico.

have the same properties and appearance. Some crystals display only one or two forms. In the cube, for example, only one form—the cube—is present. Other crystals display dozens of forms, not all of which are equally well developed. Many faces appear as small, facetlike surfaces cutting off or truncating corners and beveling edges of larger ones. On the mineral calcite, for example, crystallographers have cataloged more than seven hundred forms.

Constancy of Interfacial Angles

An early crystallographer, Nicolaus Steno (1638–1687), discovered empirically that the interfacial angles of the same face of a crystal of any mineral he studied were the same. In other words, if you measure the angle between the specific faces on a crystal of quartz, for example, you will find this angle to be constant for those two faces on *any* quartz crystal you examine.

X-ray analysis has now shown that a given mineral is characterized by a specific crystalline structure. Since the structure is constant for a mineral species, and the crystal faces always have the same relationship to this structure, the angular relationships between any two faces must likewise also be constant.

The angular relationship is not affected by the size or shape of the crystal or of the respective faces. The growth conditions of a crystal would also not affect the interfacial angles, unless the crystal's composition was markedly altered during growth.

UNDEVELOPED CRYSTALS

Specimens that might be regarded as "perfect crystals" are very rare in nature. In most instances, faces are missing, broken, or poorly developed. In some cases only a few natural faces may be present. Sometimes minerals contain one or more flat surfaces that appear to be natural faces, but that are really inherited from the shape of the cavity between other earlier-formed crystals where they grew. Not infrequently, the beginner is greatly puzzled over oddities of this kind.

Crystals that are completely bounded by faces are said to be *euhedral.* Those partly bounded are referred to as being *subhedral.* Pieces of material without natural faces, but bounded by irregular surfaces impressed upon

them by the adjoining minerals during growth, are said to be *anhedral,* that is, without any natural faces. The term *grain* is commonly applied to a small piece of material that is bounded by irregular, uneven surfaces. We speak of grains of sand, grains of salt, and grains of sugar—particles too small for the human eye to see their shapes. However, if a hand lens is used, well-formed minute crystals may be seen as constituents of some sands. Even table salt is found to consist of tiny cubes, the size of which is controlled in the refining process.

The size or shape of a mineral crystal or grain has no effect on its physical and chemical properties. When proper instruments are available, any mineral fragment can be used for identification.

CRYSTAL CLASSIFICATION

As we have seen, all crystals have symmetry that can be classified according to certain laws. Lattice types that are mathematically possible lead to six basic crystal systems. Combinations of symmetry elements about a point in space are restricted to thirty-two groups, providing thirty-two crystal classes. Earlier, we discussed crystals and lattices as abstract, mathematical entities. In the real world, of course, crystals are made up of atoms. Electrical bonding forces hold these atoms together and cause them to adopt particular spatial arrangements. These arrangements can be identified as characteristic of one or another of the crystal classes and crystal systems.

The relative lengths of crystallographic axes help to define the crystal lattices. In real crystals, relative distances become absolute distances when atoms and interatomic distances are considered. In spite of the very large number of possible combinations of atoms in crystals—considering that there are ninety-two naturally occurring chemical elements—there are still just thirty-two possible combinations of symmetry elements—planes, axes, and centers. Of these, only about half are represented by the common minerals.

Crystal Classes

Crystal classes are generally named after the form in each class that has the maximum number of faces. The names are taken from the Greek words for these geometrical solids. The number of classes varies from system to system. In the isometric system there are five; in the tetragonal system, seven; in the hexagonal system, twelve; in the orthorhombic, three; in the monoclinic, three; and in the triclinic, two.

The number of faces in a form in a given crystal class is a function of symmetry. For example, there are only two triclinic crystal classes. One of these contains only a center of symmetry, and so the maximum number of faces in any triclinic form is two. One of the triclinic classes is without symmetry, and so forms in this class consist of only one face. Triclinic crystals tend to have simple shapes. The opposite is true with the isometric system, where many symmetry elements interact to repeat a crystal face many times. Isometric forms tend to be very complex, but some of them are familiar—cube, octahedron—common geometric solids that we know from childhood. This familiarity gives the beginner a grasp of the study of crystal forms, known as *morphological crystallography.* Such a study—though beyond the scope of this book—is challenging and intriguing. It is one of the basic starting points for the study of mineralogy, and has lasting appeal to collectors of the finest mineral specimens.

MINERAL HABITS AND APPEARANCES

It is seldom that crystals grow singly and form their ideal shapes. The *habit* of a mineral is its tendency to develop into different shapes under various conditions. A given mineral species may be so influenced by its surroundings that it will have a number of different habits, depending on the different environments in which it grows. Specimens from some deposits have such characteristic habits that the type of formation in which they occur (called their *occurrence*) can be easily determined from their shape.

The terms used to describe a mineral's habits are derived from the shape of its crystal form, such as cubic, octahedral, prismatic, or tabular; or from the mineral's tendency to develop as twins; or from its structure. It may be bladed, fibrous, or micaceous, or it may even lack any particular structure, in which case it

Opposite: Crystals of calcite, hexagonal system (Cumberland, England).

Crystals in the monoclinic system: *Top:* Orthoclase (Goodsprings, Nevada). *Above:* Selenite variety of gypsum (Chihuahua, Mexico). *Right:* Azurite, partly altered to malachite (Tsumeb, South-West Africa). *Opposite:* Topaz, a mineral crystallizing in the orthorhombic system (Plateau Province, Nigeria).

may be called compact or massive. Further terms used to describe mineral habits are given in the appendix.

To the beginner, some of the variations in crystal shape may not seem to differ from the perfect, ideal crystals. Some of them however, possess the outward symmetry of a crystal system to which they do not in fact belong. Crystals of this nature are said to be distorted. When material is not added uniformly to all faces, but at a greater rate in some particular direction, the crystals lose their ideal shapes and outward physical symmetry.

The physical and chemical properties of such crystals are the same as those of the ideal crystals in every instance. Moreover, the interfacial angles are always the same, and a projection made by measuring the angles with a goniometer will quickly reveal the true symmetry of the crystal.

Distorted Crystals

The uncertain identity of a crystal is sometimes due to its distortion. Some crystals of a cubic mineral may occur as ideal equidimensional crystals in one mineral deposit, while in another they may be twice as long in one direction as in the other two; again, the cube may be much smaller in one direction than in the other two. Distortions of this nature are common for the mineral pyrite. Occasionally, with some minerals, the elongation of the cubic crystals in one direction is so extreme as to produce very slender, hairlike crystals, as seen in the variety of cuprite that is called chalcotrichite.

A quartz crystal with the usual and common faces may appear in many distorted shapes. One common distortion appears among the brilliant quartz crystals found in the Little Falls dolomite, in the southern Adirondack Mountains of upper New York State.

Crystals of minerals that crystallize in the other systems are also frequently distorted from their ideal proportions; for example, greatly elongated, needlelike crystals of tourmaline and rutile. Distortions are often considered to be the habit of the mineral.

If you take up the fascinating study of crystals, you will find it best to learn the ideal crystal forms of as many minerals as possible. When you examine a specimen, bear in mind that crystals are rarely, if ever, perfect. You can then enjoy collecting and interpreting the crystallography of distorted crystals, without being discouraged by the irregularities of forms you are sure to encounter.

Twinned Crystals and Twin Laws

The occurrence of numerous large, well-shaped crystals that are completely surrounded by smooth faces is rather uncommon for the average mineral deposit. In some localities, however, you may find a surprising number of good mineral specimens that are more than half enclosed by crystal faces. Some crystals will appear always to be intergrown in a regular way. But most of the intergrowths will be found to lack any regularity governed by crystallographic law. Most arrangements are clearly a haphazard joining, caused by the common interference among many crystals developing around each other at the same time. Minerals often make contacts with their neighbors, and because of the interferences encountered in this manner, they develop many intricate aggregates.

Those crystals that reveal some definite

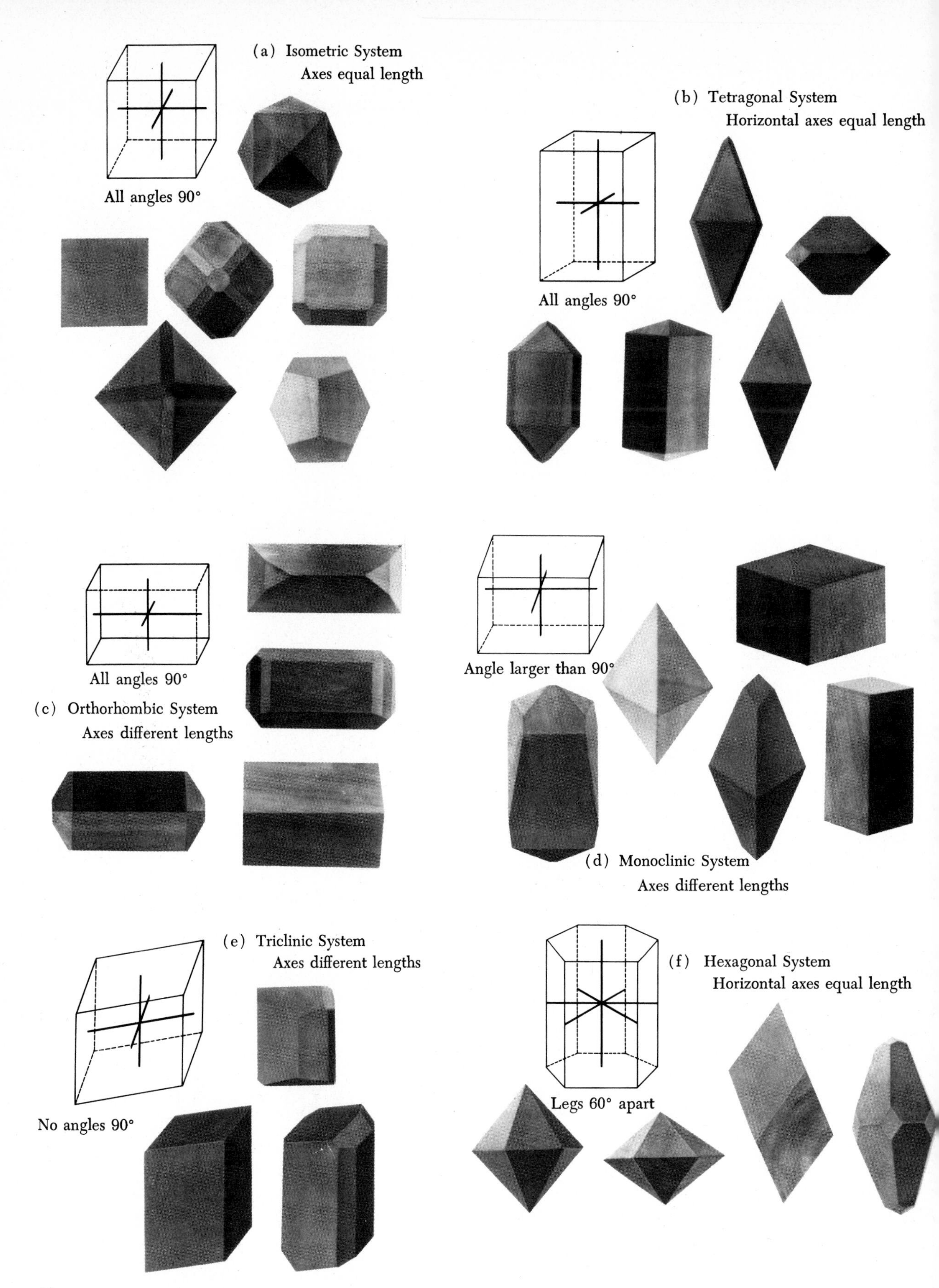
(a) Isometric System
Axes equal length
All angles 90°
(b) Tetragonal System
Horizontal axes equal length
All angles 90°
All angles 90°
(c) Orthorhombic System
Axes different lengths
Angle larger than 90°
(d) Monoclinic System
Axes different lengths
(e) Triclinic System
Axes different lengths
No angles 90°
(f) Hexagonal System
Horizontal axes equal length
Legs 60° apart

The six crystal systems, with their characteristic axes of symmetry and block models of crystals typically formed in each system.

crystallographic relationship between what appear to be the parts of two or more crystals are said to be twinned crystals. The individual parts always have a definite crystallographic orientation to each other. Such combinations always obey certain crystallographic laws, called twin laws, by which the combinations are usually known.

The *designation of twins* is based on their usual appearance. This is well illustrated by the twins of calcite and spinel. In the first instance, the spreading nature of the calcite twin has earned it the name *butterfly twin*. The spinel twin looks like an octahedron cut into two parts parallel to an octahedron face, with one of the parts rotated 180° and placed in contact with the other; it is known as a *contact twin*. The two parts of a contact twin are always mirror images of each other. In another type of twinning, the parts of the twinned crystals penetrate each other in an apparently complicated manner and are known as *penetration twins*. Staurolite frequently forms penetration twins in the shape of a cross; known as "fairy crosses," these are found in abundance in Georgia and Virginia. Another common penetration twin occurs in pyrite, forming the so-called iron cross.

The same twin law is sometimes repeated a number of times. The resulting twins are called *multiple twins*.

Twinning is frequent in the feldspars. Minerals of the plagioclase feldspar group are almost always twinned; this is shown by the prominent striations or narrow bands—like closely ruled parallel lines—that appear on certain faces of their crystals and cleavage surfaces. The crystals or mineral grains of plagioclase often consist of many thin plates or lamellas, each plate being one of a series of multiple twins. This type of repeated twinning is known as *polysynthetic twinning*, which shows up under the polarizing microscope.

In another type of multiple twinning, the twin planes may not be parallel, as they are in plagioclase feldspars, but may tend to converge about a common line, producing a circular grouping of the twin parts. If the angle between the planes is a close multiple of 360°, the resulting twins often form a circular group and are consequently referred to as *cyclic twins*. Twins of this nature often look like crystals in a higher grade of symmetry.

A simple form of cyclic twinning consists of only two crystals joined at an angle resembling a bent knee, from which it derives its name, *geniculated twin*, which is seen commonly in rutile and zircon.

In other types of twinning, selenite, a variety of gypsum, forms twins with a striking reentrant angle, which are consequently known as *fishtail twins*. Quartz twins occur in abundance and are formed according to several twin laws—the *Dauphiné law*, the *Brazil law*, and the *Japan law*, named for the areas where excellent examples have been found.

Indications of Twinning

The presence of twinning is often made apparent by several factors:

the regularity of contact or intergrowth between two crystals (or parts of crystals), other than in a repeated parallel position;

an abrupt change in the direction of the striations across an apparent symmetry plane or along an irregular boundary, producing set-off patches or striations;

the presence of reentrant angles, which occur often between twinned parts but never occur on single crystals;

a degree of symmetry apparently higher than that of the class to which the mineral belongs; and

cleavage surfaces joining at slight angular variations.

Highly complex forms of twins will, no doubt, be encountered in the field and will be found to have real determinative value. For example: the black minerals—spinel, magnetite, and chromite—are plentiful as small octahedrons in stream-bed gravels and in rocks. They can be reasonably well differentiated if there are twins among them—spinel and magnetite both twin according to the spinel law, which eliminates chromite as a possible identification. Magnetite and spinel also can be separated easily, since magnetite jumps toward the magnet and spinel does not.

Before studying twin crystals in comparison with the single forms, it is important first to be familiar with the simpler forms of each crystal class.

3. Rocks in Which Treasures Lie

The earth is a vast melting pot, in which the chemical elements—which occur naturally—are mixed and combined. The result of this process is the formation of minerals, which are the unit ingredients in nature's chemical stockroom.

Minerals may occur singly in vast deposits, such as limestone and marbles, which are made up of the mineral calcite. Usually, however, several minerals occur in the same locality. Large aggregates of minerals or large masses of a single mineral are called rocks, and rocks are thus the "home" of minerals, which may either form part of the rock itself or have formed within the rock at some later time.

The earth is dynamic, with many changes occurring constantly both on the surface and under it. The forces of destruction, such as wind, water, and ice, continually tear away at rocks exposed on the earth's surface, while internal forces melt down the residue of older rocks to create raw materials for the ceaseless cycle of rock formation.

Scientists have been able to classify rocks according to their mineralogical and chemical composition and the way in which they were formed. Geologists love to create new and interesting names, and one of the apparently confusing parts of the science of rocks, known as petrology, is the abundance of terms. But any person can remember the names of, say, one hundred friends and relatives; so a few dozen rock names, equally useful in recognition, should not be too intimidating.

There are basically three types of rocks. *Igneous* rocks form from molten material. Such rocks may form within the earth, in which case they are called *intrusive,* or on the earth's surface in association with volcanic activity, where they are known as *extrusive*. The size and shape of igneous rock bodies depend on the mode of formation and often on the composition of the rock.

At the earth's surface, rocks of all types are worn away by the process of erosion—chiefly wind, water, and ice. Bits and pieces of previously existing rocks are called sediments; these are eventually redeposited in another locality and may accumulate to vast thicknesses. The pressure created by this weight of sediments can squeeze the bottom layers to the point where they solidify into new rocks, called *sedimentary* rocks. Because of their mode of formation, many sedimentary rocks display pronounced banding or layering. Other types of sedimentary rocks result from the accumulation of chemical precipitates in oceans or lakes; this is true of nearly all limestones, for example.

Earth forces of pressure and heat can alter rocks of all types. Under such conditions, minerals may actually decompose and reform into new minerals, which are more stable under the high temperature and pressure conditions. Rocks formed in this way are called *metamorphic* rocks, whose very name signifies change A typical feature of metamorphic rocks is their increased grain size—because mineral crystals in a rock may actually be forced to coalesce—and a distinct curvature, bending, or severe twisting of preexisting rock textures.

All three rock types interact with each other. Igneous rocks are eventually exposed at the earth's surface, where they are worn down, creating sediments. Erosional forces abrade older sedimentary and metamorphic rocks as well. The new sediments are compressed into new rocks, which are in turn affected by the forces of metamorphism. Eventually all are remelted within the earth, assuring the perpetuation of the ever-changing, fascinating cycle of rock formation.

IGNEOUS ROCKS

Intrusive Igneous Rocks

A great mass of crystallized magma—or molten rock—is known as a *batholith.* It is very irregular in shape, usually somewhat elongated, and it may be hundreds of miles long and up to fifty miles or more wide. Within a batholith the rock may be quite uniform in composition, or it may vary considerably, depending upon the homogeneity of the magma and the amount and kind of wall rock that was melted and incorporated into the magma during its contact with adjoining rocks. The texture—that is, grain size—and the grain arrangements are usually rather uniform. The various parts of the main body of a batholith are in general much alike in appearance and texture.

In the field, igneous rocks appear in a great

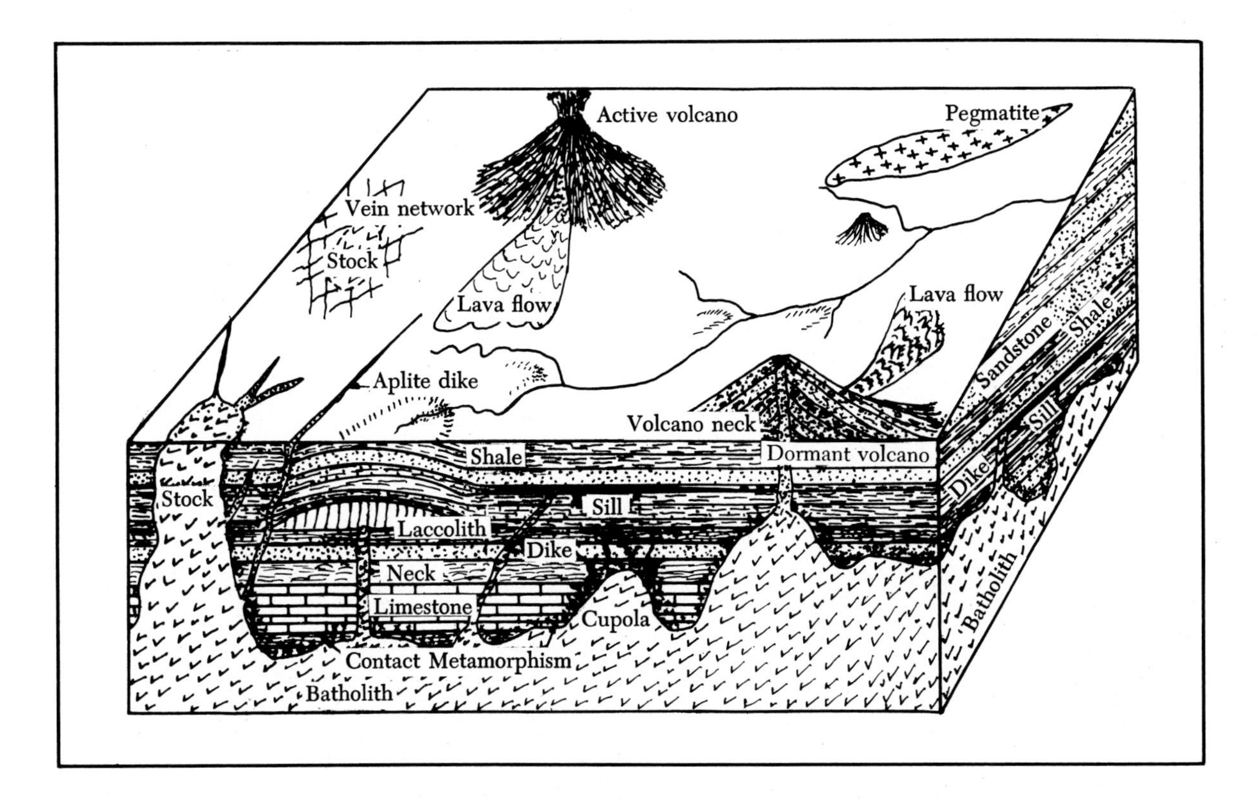

variety of forms and shapes. The exact shape of any body of igneous material may be largely concealed by the enclosing rock; however, the majority of occurrences are clearly visible. The common types of igneous rocks occur in close association with the main mass or batholith, though it is unusual for all types to occur in a single location. Among the larger formations are the so-called *stocks,* which represent the upward extensions of great knobs of rock. Surrounding the stocks is usually a system or network of *veins* or *dikes* penetrating the surrounding rocks. Such networks usually contain the same kind of rock as the batholith, although their texture may be either coarser or finer. The smaller dikes may consist of a very fine-grained rock, called *aplite* because of its sugary texture and not because of its mineral composition, which is usually close to that of the main igneous body.

The upper surface of a batholith is not flat or smooth; instead, it is exceedingly uneven, with dikes and irregular masses of the batholith extending upward into the enclosing rocks. Masses smaller than stocks are called *cupolas,* which, like stocks, may be surrounded by a system of veins or dikes. Such vein systems oc-

Preceding pages: The Devil's Tower, a unique formation of huge basalt columns in eastern Wyoming. *Above:* A batholith—an igenous rock formed from the molten state—and the various types of rock formations that may accompany it. Volcanoes, stocks, laccoliths, dikes, sills, and cupolas feed off from the batholith. Other, overlying formations include sedimentary rocks —shale and sandstone—and metamorphic rocks.

casionally contain mineral deposits of economic value.

Occasionally some of the intruded magma may have spread out along the bedding planes of overlying sediments, to form extensive, flat, interbedded layers. These layers are called *sills,* and in some rock quarries they often look much like another layer of sediments. Dikes are structures that cut across preexisting layered rocks, rather than conforming with them. They are common throughout the world in the older igneous and sedimentary rocks. Dikes are not always the feeding channels for other rock forms; they may end at the furthest extremity of a fracture.

Dikes and sills, except for their physical relationship to the enclosing rocks, are essentially the same—that is, both are irregular plates of consolidated magma. Each has been injected or intruded into relatively cool rocks and has solidified rather quickly, with shrinkage along the length and breadth of the sheetlike surfaces where there is contact with the surrounding rocks. The shrinkage continues, causing polygonal cracks in the solidified material. These cracks extend through the rocks to form columns that stand at right angles to the enclosing rock walls. Within a partially eroded dike, these columns may resemble a staircase; such a structure is known in German as *Treppe,* which has become corrupted in English to "trap"—hence the name "trap rock."

Occasionally, when an intrusive mass of magma reaches the upper strata of sediments, it pushes the layers up into a *laccolith*—a dome shaped somewhat like a toadstool, although often irregular. The feeding channel for this form is called a neck, if it is nearly cylindrical, or a dike, if it is a plane surface.

Extrusive Igneous Rocks

The most spectacular display accompanying

the movement of vast bodies of magma is volcanic activity. Volcanoes come into being when a magma dissolves the overlying rocks until it reaches the surface. When the high pressures of dissolved gases are no longer confined, they are free to escape violently into the atmosphere, carrying along vast quantities of red-hot magma and some of the adjacent wall rock. The ejected material falls back around the volcanic vent or crater, solidifying into a cone-shaped mass of fragmental rubble called a volcano, inside of which is a feeding channel, the volcanic neck.

Volcanoes have been spectacular sights in mountainous areas around the world. A notable circle of volcanic activity—called the ring of fire—borders the Pacific Ocean, extending from the southern Andes through Mexico, along the Rocky Mountains to Alaska, thence following the Aleutian Islands and down the Asian coast through Japan to the islands of the South Pacific, and east again across the Pacific Ocean. Volcanic activity is not a phenomenon of the past alone; it is still a vital and operating process in the earth. Within the past hundred years we have witnessed the formation of two new volcanoes, Parícutin in Mexico and Surtsey in the ocean near Iceland. There will, of course, be more, but it is hard to predict exactly when and where—possibly in the Rocky Mountains, or around Iceland, in the mid-Pacific near the Hawaiian Islands, or in the southwest Pacific.

Volcanic eruptions are frequently accompanied by lava flows, which may extend several miles from their source.

Probably the most spectacular processes on earth are eruptions from fissures—cracks in the earth's crust—through which an unbelievable amount of fiery, red-hot basaltic magma wells up from the depths of the earth and flows away over vast areas on the surface. Millions of years ago in Washington and Oregon, 200,000 square miles were covered with molten basaltic lava erupting from fissures, reaching a depth of four thousand feet. A similar flow occurred in northwestern India on the Deccan Plateau, where basaltic lava covered hundreds of square miles and reached a depth of six thousand feet in some places.

Once in contact with the air, the lava or molten rock loses heat quickly, and a rock crust forms over it. If the crust is securely anchored, the fluid lava may sometimes drain away, leaving channels or open tubes—known as lava tubes—beneath the surface. These structures are not uncommon in lava fields but are usually

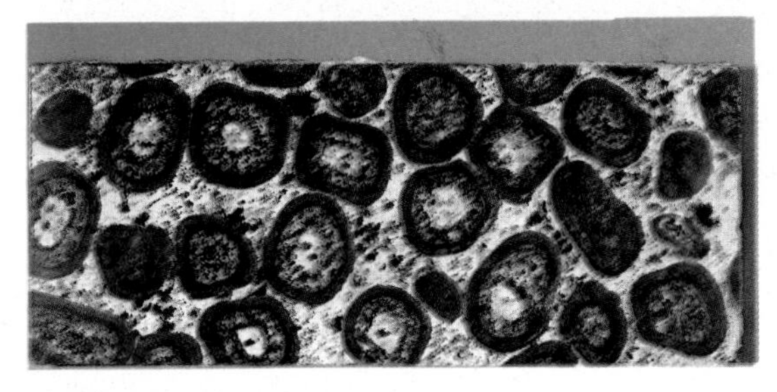

Opposite: Granite intrusion (tan) in sedimentary rock (gray), Wyoming. *Above:* A basic dike, with horizontal joints formed by shrinkage (Wyoming). *Top right:* Trachyte porphyry, a rock with large crystals (California). *Above right:* Orbicular granite (Kostfors, Sweden).

hidden from sight. Sometimes their roofs collapse, leaving gullies or valleys through the lava field. This is the origin of Craters of the Moon National Monument in southern Idaho.

The quick cooling of the surface of extrusive rocks may cause shrinkage cracks at the outer contacts. These polygonal, cracked areas extend toward the base, forming the characteristic basaltic columns that occur prominently in many lava flows. An example of this formation may be seen at Devil's Postpile National Monument in Madera County, California.

Since volcanic cones are composed of loose and poorly compacted fragments, they are quickly eroded to the level of the surrounding land; but the central upright crystalline core may remain as a conspicuous landmark for many years. An excellent core or neck may be seen near Bernal, Mexico; several occur in the Four Corners area of Colorado, Utah, New Mexico, and Arizona; and others are in southwestern South Dakota. In South Africa a few of the ancient volcanoes and their central cores have been eroded to the surface level.

Volcanoes explode because of the confined, highly heated gases they contain. Such explosions may be periodic. The more recent violent and spectacular eruptions were Mount Pelée on the island of Martinique in 1902 and the extremely violent explosion of Mount Krakatoa in the East Indies in 1883. The latter blew into the air seven cubic miles of earth and rocks, some of which flew to a height of seventeen miles.

Above: A volcanic bomb, from near Mt. Trumbull, Ariz.
Below: Weathering can create fantastic shapes, like this balanced rock (Chiricahua Nat. Monument, Ariz.).

Mineral Composition of Igneous Rocks

To determine the names of igneous rocks, we need to identify a relatively few rock-forming minerals. Chief among these are quartz, orthoclase feldspar, biotite, hornblende, augite, plagioclase feldspar, and olivine. About a dozen others occur in very small amounts as accessory minerals. It is surprising that the rocks of the earth's crust, down to ten miles below the surface, are composed of such a small number of minerals.

The igneous rocks are the most abundant of all rocks to a depth of ten miles; below this, all rocks are probably igneous. Although about ninety-two chemical elements in the earth's crust have been identified, only eight make up

97.14 percent of the rocks. These very important rock-forming elements are reported as oxides: silica (SiO_2), 70 percent; aluminum oxide (Al_2O_3), 15 percent; ferric oxide (Fe_2O_3) and ferrous oxide (FeO), 2.6 percent; magnesium oxide (MgO), 0.64 percent; calcium oxide (CaO), 1.7 percent; sodium oxide (Na_2O), 3.2 percent; and potassium oxide (K_2O), 4 percent. The average mineral composition of the igneous rocks in the earth's crust to a depth of ten miles is:

Quartz	12 %
Feldspars	59 %
Pyroxene and hornblende	16.8%
Mica	3.8%
Accessory minerals	8.4%

If we were to stand on the brink of one of the Hawaiian volcanoes at the time of volcanic activity, we would have a first-hand look at magma, or molten rock. The magmas, which yield the different kinds of rocks, are solutions of various compositions at temperatures ranging from around 500° C to 900° C. Those richer in silica (SiO_2) crystallize at higher temperatures than those rich in alkalies (potassium and sodium). Magmas originate at considerable depths beneath the surface, under conditions not entirely understood and in volumes up to many thousands of cubic miles. As the magma cools, it reaches a point on the temperature scale where one of its minerals crystallizes. As the temperature is continually lowered, other minerals will, in succession, become saturated and form solids, until the entire mass of magma has solidified. The solution does not always crystallize completely, for the magmas possess a varying amount of gases that do not form solids. When the temperature drop is rapid, there is insufficient time for crystallization to occur throughout the magma, resulting in the formation of igneous glass or only partial crystallization of the magma.

The larger the mass of magma is, the more slowly it cools, and because of the slow cooling, the entire volume may be completely crystallized and is sometimes coarse-grained. We notice the opposite effect when magma is injected into fractures in cold rocks, where the rapid cooling allows only partial crystallization and the formation of a natural glass.

One of the important factors controlling the rate of crystallization is the amount of mineralizers and low-melting minerals any magma contains. Mineralizers include such substances as water, fluorine, boron, chlorine, and also car-

An exposed sill in Yellowstone Park, Wyoming. The magma has pushed in between sedimentary layers to produce a layer of igneous rock. The vertical columns formed when the hot magma shrank as it cooled.

Two high-silica, crystalline rocks found in batholiths and laccoliths. *Top:* Coarse-grained granite containing biotite, from Peterhead, Scotland. *Above:* Fine-grained hornblende syenite, containing little or no quartz, from Vermont.

bon dioxide. Any minerals that act as fluxes or solvents are also called mineralizers.

Rocks are characterized by their component minerals, as well as by the size of the mineral crystals and their pattern of intergrowth. Such patterns are called *textures.*

The most common texture is granitoid or granular, which varies from fine-grained to coarse, or from barely visible grains to those the size of a thumbnail. Granular rocks are found in all categories of igneous rocks. In a type of coarse-grained granite known as pegmatite, the size of the individual mineral grains varies from minute grains to those of fifty or a hundred tons! These coarse-grained minerals are restricted almost entirely to the granite pegmatites.

In magmas, often one or two mineral species become saturated at a higher temperature and develop euhedral (full-faced) crystals up to a half inch in size prior to the crystallization of the remaining solution, which forms the ground mass or matrix for the earlier crystals. The large crystals are called phenocrysts, and the rock is known as porphyry.

Classification and Description of the Common Igneous Rocks

Igneous rocks are commonly classified according to their chemical content, mineral composition, and grain size. They are arranged in ten groups or families, which we may define as distinct groups although in nature they are gradated from one group to the next. This arrangement begins at the top with the most "acid" rocks—that is, those highest in silica—and extends to the most basic ones below. The name of any rock with a specific composition is determined by the grain size of the constituent minerals. If these minerals can be identified at sight or with a hand lens, identification and naming of the rock is simple. In the case of fine-grained (felsitic) or glassy rocks, which have no visible grains, identification may be difficult. Sometimes clues are apparent that aid in identification; otherwise laboratory tests must be used.

Igneous rocks are also classified according to the type of feldspar they contain, and sometimes according to the presence of certain accessory minerals, which are distinctive minerals that may be present in small amounts. Acidic rocks generally contain feldspars rich in sodium and potassium. Basic rocks contain feldspars rich in calcium. Group names are based on the names of rocks associated with batholiths, for rocks of any given compositional range.

Granite-Rhyolite

Granite consists of orthoclase feldspar and one or more of the minerals biotite, hornblende, and augite, in various combinations with quartz and orthoclase feldspar (the latter two are the essential constituents in granites). Granite is *holocrystalline*—that is, completely crystallized and having a more or less uniform grain size. Granites are common all over the world. They vary in grain size from fine to coarse and in color from white to very dark; pink, reddish, and white granites are common. In some localities granites have large crystals or grains of feldspar that are many times larger than the ground mass; such rocks are known as *granite porphyries.* When the granite magma has cooled very fast, the crystallization has been so rapid that the grain size is too small for the individual grains to be distinguished even with a hand lens. Rocks of this nature in the acid and intermediate rocks are said to be *felsites*—a general term—and a granite felsite is also known as a *rhyolite.* As these rocks are rich in silica, it is not unusual to be able to identify a few large grains of quartz that stand out clearly in the ground mass.

When the crystals are prominent, the rock becomes a *rhyolite porphyry* or a *quartz porphyry.* When the texture becomes glassy, the identity of the acid rocks cannot be determined except by chemical analyses. These glassy rocks are known as *obsidian, perlite, pumice,* or *pitchstone,* depending largely on their physical appearance. The granite-rhyolite rocks contain an average of about 70 percent silica, in contrast to the basic rocks—augite-peridotites—which contain around 40 percent silica.

Among the many variations in granite structures, the orbicular phase is one of the most interesting. It consists of scattered orbs

Top: The slower a magma cools, the larger the crystals that may form, as in this granite porphyry from New Hampshire. *Above:* A rapidly cooled magma has no chance to form crystals; hence, its texture is smooth, like this obsidian from Guadalajara, Mexico.

CLASSIFICATION OF THE COMMON IGNEOUS ROCKS

<table>
<tr><td colspan="4">OCCURENCE</td><td>Beds</td><td>Dikes, bosses, injections</td><td colspan="2">Batholiths, laccoliths</td><td colspan="2">Dikes, intrusive sheets, laccoliths</td><td>Surface flows</td><td>Crusts, surface flows</td></tr>
<tr><td colspan="4">TEXTURE</td><td rowspan="2">Fragmental</td><td rowspan="2">Coarse-grained</td><td colspan="2" rowspan="2">Granitoid</td><td rowspan="2">Porphyritic, phenocrysts predominate</td><td rowspan="2">Porphyritic, phenocrysts prominent</td><td rowspan="2">Cellular, glassy, or felsitic; phenocrysts few</td><td rowspan="2">Glassy</td></tr>
<tr><td>GROUP NAME</td><td colspan="3">MINERAL COMPOSITION</td></tr>
<tr><td>Granite-rhyolite</td><td rowspan="3">Ortho-clase feldspars</td><td rowspan="3">Biotite, horn-blende, augite</td><td>+Quartz</td><td>Rhyolite tuffs and breccias</td><td>Pegmatite</td><td colspan="2">Granite</td><td>Granite porphyry</td><td>Rhyolite porphyry (quartz porphyry)</td><td>Rhyolite (felsite)</td><td rowspan="3">Acid glasses, perlite, acid obsidian, pumice, pitchstone</td></tr>
<tr><td>Syenite-trachyte</td><td>−Quartz</td><td>Trachylite tuffs and breccias</td><td>Pegmatite</td><td colspan="2">Syenite</td><td>Syenite porphyry</td><td>Trachyte porphyry</td><td>Trachyte (felsite)</td></tr>
<tr><td>Nepheline-phonolite</td><td>Nepheline or leucite</td><td>Phonolite tuffs and breccias</td><td>Pegmatite</td><td colspan="2">Nepheline syenite</td><td>Nepheline-syenite porphyry</td><td>Phonolite porphyry</td><td>Phonolite (rare), leucite rocks (very rare)</td></tr>
<tr><td>Quartz diorite-dacite</td><td rowspan="4">Plagio-clase feldspars</td><td rowspan="2">Biotite, horn-blende</td><td>+Quartz</td><td>Dacite tuffs and breccias</td><td rowspan="6">Pegmatites less frequent in the more basic rocks</td><td colspan="2">Quartz diorite</td><td>Quartz-diorite porphyry</td><td>Dacite porphyry</td><td>Dacite (felsite)</td><td rowspan="4">Andesite (basic) obsidian</td></tr>
<tr><td>Diorite-andesite</td><td>−Quartz</td><td rowspan="2">Andesite tuffs and breccias</td><td colspan="2">Diorite</td><td>Diorite porphyry</td><td>Andesite porphyry</td><td>Andesite (felsite)</td></tr>
<tr><td>Gabbro</td><td rowspan="2">Pyroxene</td><td>−Olivine</td><td>Gabbro</td><td rowspan="2">Dia-base</td><td>Gabbro porphyry</td><td>Augitite-andesite porphyry</td><td>Augite andesite</td></tr>
<tr><td>Olivine gabbro</td><td>+Olivine</td><td rowspan="3">Basalt tuffs and breccias</td><td>Olivine gabbro</td><td>Olivine gabbro porphyry</td><td>Basalt porphyry</td><td>Basalt</td></tr>
<tr><td>Pyroxenite</td><td rowspan="2">No feldspars</td><td rowspan="2">Augite, horn-blende, biotite</td><td>−Olivine</td><td colspan="2">Pyroxenite</td><td>Pyroxenite porphyry</td><td>Augitite porphyry</td><td>Augitite</td><td rowspan="2">Basic glasses, scoriae, trachylytes, basalt obsidian</td></tr>
<tr><td>Peridotite</td><td>+Olivine</td><td colspan="2">Peridotite</td><td>Peridotite porphyry</td><td>Limburgite porphyry</td><td>Limburgite</td></tr>
<tr><td colspan="12">Ultra basic rocks: Basic segregations in normal magmas, meteorites, water, and ice.</td></tr>
</table>

Top left: Felsite porphyry (Bannockburn, Ont.), with large crystals formed deep in the earth, grains formed by rapid cooling near the surface. *Top right:* Quartz diorite, from intrusive igneous formations (Riverside Co., Cal.). *Above left:* Olivine gabbro, a coarse-grained igneous rock (Wichita Mts., Okla.). *Above right:* Andesite porphyry, related to diorite but with larger crystals (San Bernardino Co., Cal.).

of banded minerals in a ground mass of granite. The orbs have a center of feldspar and quartz surrounded by alternating narrow bands of biotite and hornblende, with succeeding bands of feldspar and quartz.

Syenite-Trachyte

With the decrease of quartz to zero in the granite-rhyolite group, the rocks enter the syenite-trachyte group. In field work the task of determining the absence of all quartz is often very difficult, as a small amount of quartz grains can be effectively concealed. It is quite possible for a field worker to be unable to establish a sharp line between the two families without the use of the petrographic microscope. The term *trachyte* refers to a subparallel arrangement of the small lath-shaped feldspar crystals. In this group the holocrystalline rock is the *syenite*.

Nepheline-Phonolite

With the decreasing amount of silica and increasing amount of sodium in the magmas, quartz is no longer free to form, and other minerals appear. Nepheline in some ways resembles feldspar and is often referred to as a *feldspathoid*. Both nepheline and leucite occur in the low-silica magmas rich in sodium and potassium. This group of rocks is similar to the syenite family, but the holocrystalline rock is designated *nepheline syenite*, while the felsitic phase is known as *phonolite*—in reference to the clear, ringing sound produced by thin, suspended slabs when struck with a hammer. As nepheline and leucite are not common minerals, the nepheline syenites and phonolites are among the rarer rocks.

Quartz Diorite-Dacite

In the granite-syenite group, the chief feldspar is orthoclase, with occasionally a minor amount of plagioclase present as an accessory mineral. When plagioclase becomes the dominant feldspar together with quartz, the rock becomes a *quartz diorite*; the presence of plagioclase as the chief feldspar reveals a decrease in the amount of silica and an increased basic nature of the rock. This change is usually accompanied by an increase in the dark accessory min-

Above: Photomicrograph of diabase. The feldspars form first, in lathlike crystals, and dark minerals form around them (York Co., Pa.).

erals, which darkens the average color of the rock. The felsitic rock—*dacite*—is dark, too fine-grained for visual determination. The glassy rock in this group is andesite obsidian, which in general lacks the pronounced vitreous luster of the acid glasses. The diorites in the field are distinguished from the granites by the detection of twinning striations on the average faces of the plagioclase feldspars; they may be seen with a hand lens.

Diorite-Andesite

Diorite is similar to syenite in its mineral content and lack of quartz. In diorite, however, the feldspar is plagioclase rather than orthoclase. In addition, the overall appearance of the rock is darker than syenite because the dark accessory minerals are more prominent. *Andesite porphyry* is a felsitic—fine-grained—rock containing phenocrysts; its name is derived from the Andes Mountains, where this rock type is abundant.

Gabbro, Olivine Gabbro

In diorites and andesites, the chief accessory minerals are biotite and hornblende, while augite predominates in the gabbro and olivine gabbro. The feldspars are farther toward the basic end of the series. The pyroxenes are the most prominent mineral species in the gabbros, while the feldspars are usually subordinate and occupy the interstitial spaces between the dark minerals. However, there are instances where the order is reversed, and then the pyroxene minerals occupy the spaces around the feldspars, which appear as prominent, twinned, lath-shaped grains. These rocks are known as *diabases* and are common in large dikes, laccoliths, and the central parts of thick flows. The most basic gabbro is the first to contain the basic mineral olivine (peridot), and in the felsitic phase it constitutes the true basalt. However, the very basic felsitic rocks are even more difficult to distinguish than the acid ones; hence, in the field the term basalt is frequently used to refer to similar black, basic, fine-grained rocks.

Pyroxenite, Peridotite

These two groups are similar in that the plagioclase feldspars have dropped in amount to the

Top: Diabase from Ontario, Canada. *Center:* Kimberlite, a periodotite fragmental rock (Premier Mine, S. Africa). *Bottom:* Cumberlandite, a basic rock containing magnetite and ilmenite (Rhode Island).

position of accessory minerals. The pyroxenite, as its name implies, consists essentially of one or more of the pyroxene minerals. The accessory minerals are likewise very basic and include magnetite, ilmenite, apatite, chromite, corundum, spinel, garnet, pyrrhotite, pentlandite, and the basic feldspars.

The peridotite rocks, on the other hand, contain olivine (peridot), often in large amounts. Peridotite is usually found in small dikes in many parts of the world. It also occurs as a *breccia* (fragmental rock) in volcanic necks, where it is known as kimberlite; this is the rock in which diamond occurs as a rare accessory mineral. Other accessory minerals are magnetite, ilmenite, chromite, pyrrhotite, spinel, garnet, apatite, corundum, rutile, serpentine, chrysotile, chlorite, epidote, and some carbonates, such as calcite and siderite.

A rock that consists chiefly of olivine with minor amounts of chromite is known as a *dunite*.

Ultra Basic Rocks

The very basic rocks consist essentially of segregations which are comprised largely of iron-rich minerals, such as magnetite, ilmenite, pyrrhotite, pyrite, and hematite accompanied by chromite, corundum, rutile, and the minerals of the pyroxenites and peridotites. Their texture is usually granitoid with magnetite or magnetite-ilmenite minerals the most common. Cumberlandite from Rhode Island is a member of this group and contains 50 percent of ore minerals of iron and titanium.

Ice and Water

It may seem strange to classify ice and water as igneous rocks. But water is a natural, inorganic fluid present in magmas, and hence it is a mineral; in the oceans, it has many other mineral molecules in solution. When the temperature drops low enough, water solidifies and yields a rock—ice—which persistently covers extensive areas in the polar regions of the earth.

Mineral Deposits in Igneous Rocks

Igneous magmas are the original sources of all minerals. However, with the development of sedimentary and metamorphic rocks, not all present-day mineral deposits can be ascribed directly to igneous magmas. Nevertheless, there are many types of mineral deposits that are closely associated with igneous rocks or occur in their immediate vicinity. A few outstanding ones may be mentioned.

Diamonds originate in very basic igneous peridotite rocks where the element carbon is incorporated, as in the peridotite volcanic necks of Africa. Chromite deposits are also closely associated with basic rocks, as in the norites in the Union of South Africa and the great basic dike in southern Rhodesia. Rich copper-nickel minerals, including the platinum metals, are closely associated with norites and other basic rocks in the Sudbury, Ontario, basin. The iron mines in southeastern Pennsylvania are contact deposits associated with the diabase intrusions of the area.

In the southwestern part of the United States, very large copper deposits are disseminated throughout porphyritic rocks known as monzonite and quartz monzonite. The content of these rocks is characterized by an equal amount of orthoclase and plagioclase feldspar; in our classification they would be placed between the nepheline syenite and quartz diorite. It is interesting to note that granite, though it occurs in vast quantities in the earth's crust, has no characteristic ore deposits associated with it. However, the rich gold veins of Canada and the western United States are extremely differentiated phases of quartz porphyry and trachyte porphyry.

When a highly heated magma comes in contact with other rocks—chiefly sedimentary rocks and especially rocks rich in calcium carbonate—it frequently effects a profound change in the sedimentary rocks. Often there are zones around the igneous rock mass in which large quantities of new minerals are introduced from the magma, including the economically important copper, iron, lead, and zinc. Minerals commonly in these contact metamorphic deposits are garnet, wollastonite, epidote, tremolite, diopside, ziosite, vesuvianite, and quartz, and sometimes axinite, tourmaline, fluorite, scapolite, and danburite.

Commercial Use of Igneous Rock

Igneous rocks were at one time used as di-

Top: Meteorite from Central Australia reveals intricate design. *Center:* Pebble conglomerate, a detrital sedimentary rock (Eng.). *Bottom:* Arkose, a sediment formed from disintegrated granite or gneiss (Mass.).

Top: Two tektites, of natural glass, found in Indochina. *Center:* Coquina, a sedimentary rock containing shells (Florida). *Bottom:* Potash salts from New Mexico, deposited by evaporation of saline bodies of water.

mension stones in the erection of large buildings, but in recent years this practice has diminished because the cost of quarrying, cutting, and laying the stones is so high. Only public buildings and large commercial structures can afford such expense. There is still a market for the better-quality granite for use as statues and monuments; here too the cost is noticeable by the relatively small size of these stones. Because of their superior hardness and toughness, the principal use for igneous rocks is in highway construction, where they are crushed and laid as a base under the surface. Only long transportation costs weigh against the igneous rocks in favor of limestone, sandstone, and metamorphic rocks.

METEORITES

Meteorites are extraterrestrial objects that have been falling to earth from outer space for billions of years. Although there are frequent arrivals every year, most meteorites are very small and are melted or disintegrated by the frictional heat generated when they enter the earth's atmosphere.

There are, however, many instances when the more massive pieces survive the hazards of entering the atmosphere and come to rest in the earth at shallow depths. Those that have been found have been located accidentally; however, a few have been observed while falling or have been located by following the path of the meteor.

Of the few meteorites that have left unusual disturbances where they hit the earth, the most famous is the Meteor Crater in Arizona. Another large impact site is the Chubb Crater in northern Quebec. Iron-rich meteorites are not abundant here; but, in contrast, tons of scattered meteorites have been gathered up, by means of various magnetic devices, in the area surrounding the huge crater in Arizona.

Meteorites vary considerably in their mineralogical combinations; the commonest and most resistant to weathering are the iron meteorites, consisting mostly of iron and 5 to 10 percent nickel. Meteorites come to the earth in sizes ranging from large masses to fine dust. The iron meteorites are among the larger pieces of extraterrestrial material on record. One of the largest is the Grootfontein iron meteorite found near Grootfontein, South-West Africa, estimated to weigh over 100 tons. Another large one, weighing 37½ tons, was brought by Admiral Peary from Cape York, Greenland, and is now in the Hayden Planetarium in New York City. The smallest ones weigh in at only a small fraction of an ounce. To the amateur, one of the most fascinating

features of the iron meteorites is the intricate design revealed in some pieces by polishing a fresh, flat surface and then etching it with acid; the resulting patterns are the so-called Widmanstätten figures.

In the early 1800s, an attempt was made to forge some swords from meteoric iron; but these irons do not contain carbon, and it is not likely that they would maintain a good cutting edge. In the prehistoric earthworks of Hamilton County, Ohio, a few artifacts of meteoric iron were found.

The second group of meteorites is the siderolites, including the pallasites, which are stony-iron meteorites containing nodules of iron, calcium, and magnesium minerals. These meteorites disintegrate relatively easily, and the small pieces soon enter the soil.

Stony meteorites, known as aerolites—the least common group—consist of basic aluminum, calcium, and magnesium minerals, which decompose and enter the soil after a few thousand years. Aerolites are too similar to common terrestrial rock to be readily distinguished; a petrographic examination is usually required to establish their identity. The largest stony meteorite, reported from Estacado, Texas, weighed 638 pounds.

A look at some of the minerals of meteorites will quickly indicate the very basic nature of these bits of disrupted planet: olivine, apatite, labradorite, graphite, diamond, enstatite, augite, anorthite, magnetite, chromite, pyrite, and pyrrhotite, among others. The stony-iron meteorites—pallasites or siderolites—consist of grains or small nodules of pyroxenes or peridotites surrounded by meteoric iron.

TEKTITES

Tektites are glass objects of unknown origin.

Opposite left: Oölitic hematite, a sedimentary rock containing iron, from Clinton, New York. *Opposite right:* Cannel coal from Kentucky, formed mainly by consolidation of fern spores. *Above:* A geode from Keokuk, Iowa, containing well-formed crystals of calcite.

Their most common form is a flattened ellipsoid. Their composition is high in silica—73 percent—with 13 percent alumina, 5 percent ferrous iron, 2 percent calcium oxide, and lesser percentages of other common rock components. The largest pieces weigh as much as twenty-eight pounds.

In an effort to determine the origin of tektites, a few investigators advocate a lunar source, while others support the theory that they originate on earth. The analysis given for tektites is close to that of the more acid granites, which at once throws doubt on a lunar origin, as the lunar rocks are far more basic.

SEDIMENTARY ROCKS

Sedimentary rocks are made of debris. The forces of weathering, including wind, rain, ice, and plant action, readily affect many minerals in rocks formed at high temperatures and pressures within the earth. Olivine, feldspars, and other minerals are decomposed, and chemical breakdown particles are carried away in solution. Resistant minerals are pulverized, broken down, and transported by streams and wind.

Eventually all this rock debris and chemical waste is redeposited, in a stream bed, a lake, or the ocean. Deposition from water occurs by settling, resulting in a separation of larger and smaller particles. Rocks formed from such deposits may therefore look stratified, or banded. The hardening or compression of masses of rock debris, called *sediments,* eventually creates new rocks. Another name for debris is "detritus," and sedimentary rocks formed from particles of various sizes are called *detrital rocks.*

Rocks may also form when breakdown products of decomposed rocks that are carried in solution are forçed out of solution. Such ejected materials are called *precipitates,* and the process is the primary origin of the rock limestone.

Detrital Rocks

The most common rocks seen by the traveler as he tours the various parts of the world are the sedimentary ones. The commonest sedimentary rocks are the detrital rocks, which are built up by the accumulation of the disintegrated and decomposed parts of previously existing rocks of all sorts.

In the weathering of a granite, there are always some unaltered detrital products. The most resistant are the quartz grains and a few of the accessory minerals, which are freed from the rock and remain as unaltered grains for millions of years. The feldspars are decomposed with the release of hydrated silica; clay particles are formed, and potassium and sodium salts move away in solution. The minerals biotite and muscovite are very resistant to decomposition, but they are easily reduced to small pieces—because of their eminent and easy cleavage—and are then transported by water or wind. Such minerals as the amphiboles and pyroxenes of the more basic rocks break down to hydrated silica, green clay minerals, serpentine, and a minor amount of soluble salts. The accessory minerals—magnetite, zircon, garnet, apatite, tourmaline, and titanite —are resistant to weathering and move in water currents along with the quartz grains to the sea or to inland basins, which are the usual deposition areas for most detrital matter.

The detrital sediments and their consolidated rocks are generally named for the kind and size of the individual pieces. Rounded pieces of rock larger than 256 mm (10 inches) in diameter are called boulders, and the consolidated rock they form is a boulder conglomerate. The composition of the pieces is immaterial; they may be rounded chunks of any mineral or rock. Pieces from 4 mm to 64 mm (2½ inches) when consolidated are called pebble conglomerate.

The grains of sands vary in size from $1/16$ mm to 2 mm, and, accordingly, the sandstones they form are described as fine-, medium-, or coarse-grained. The stone is formed by consolidation of the loose material under pressure from overlying sediments; the pore space is cemented by percolating waters carrying one or more of the compounds of silica, carbonate, and iron, which crystallize in the open spaces and cement the grains into solid rock.

A more unusual rock is coquina, which consists of cemented fragments of the shells of marine animals; it occurs locally along elevated marine beaches, as in Florida and along

CLASSIFICATION OF COMMON DETRITAL SEDIMENTARY ROCKS

Particle Size	Name of Particle	Name of Sediment	Name of Consolidated Rock
Larger than 256 mm	Boulder	Coarse gravel	Boulder conglomerate
256–64 mm	Cobble		
64–4 mm	Pebble	Pebble gravel	Pebble conglomerate
4–2 mm	Granule	Coarse-grained sand	Coarse-grained sandstone
2–1 mm	Very coarse sand grain		
1–1/2 mm	Coarse sand grain	Medium-grained sand	Medium-grained sandstone
1/2–1/4 mm	Medium sand grain		
1/4–1/8 mm	Fine sand grain	Fine-grained sand	Fine-grained sandstone
1/8–1/16 mm	Very fine sand grain		
1/16–1/256 mm	Silt particle	Clay (mud)	Shale
1/256–smaller	Clay particle		
Marine shells, usually fragmental		Shell sand or gravel	Coquina
Small to angular rock fragments		Talus	Talus breccia
Feldspar-rich sand and gravel		Arkosic sand or gravel	Arkose
Feldspathic sand plus black or gray grains		Graywacke sand	Graywacke
Ejected volcanic debris—ash, cinder, volcanic breccia (large angular pieces)			
Wind-blown sand and dust		Sand dunes and loess	Sandstone, shale

the Chilean coast. Talus breccias are seldom preserved and consolidated; however, arkose and graywacke sediments are sometimes consolidated. They are not common except in a few areas.

Silts and clays are the basic ingredients for consolidated *shales*. The consolidation of fine detrital sediments is effected by pressure from overlying sediments, which expels the water and brings the particles closer together. This results in recrystallization, during which the minute particles are dissolved and the material is added to the larger pieces, forming an interlocking mosaic of grains.

Chemical Precipitates

Among the chemically precipitated rocks, the calcium and magnesium carbonates are most important, for they form extensive sedimentary deposits. Many of these *limestone* deposits are rich in fossil remains, which often make up a considerable part of the formations. Some limestones contain high percentages of clay and sand. Many shales and sandstones are likewise rich in calcium carbonate; hence, there are *argillaceous* (shaly) and *arenaceous* (sandy) limestones, as well as *calcareous* shales and sandstones. A limestone can be readily identified by the scratch method; it is easily scratched and leaves a white powder. Also, the stone will effervesce if a drop of 10-percent cold hydrochloric acid solution is dropped on it. Dolomites are also easily scratched, but only the powder will effervesce readily in cold, dilute acid.

The carbonate rocks are used extensively in building construction. Large amounts are employed in metallurgical operations as a flux and in making road gravel.

The sudden or gradual sinking of a land area may result in the creation of a lake or inland sea. Streams flow into the sea, carrying in solution chemicals formed by rock decomposition. If the sea has no outlet, the chemicals cannot be removed and gradually become more and more concentrated. Evaporation of the

Top: A handsome septarium with calcite crystals, from Utah. *Above:* Quartzite from Alberta, Canada, a dense metamorphic rock composed of quartz. *Opposite:* "Pyrite sun," a concretion of pyrite, from Illinois.

water causes further concentration. Eventually, the water in the sea becomes saline, sometimes extremely so, as in the Dead Sea in Israel, the Salton Sea in California, and the Great Salt Lake in Utah.

If the salt concentration becomes too great, dissolved material is forced out of solution and is deposited on the sea or lake bottom. The mineral to crystallize first is a small amount of hematite, followed by precipitates of gypsum, halite, and anhydrite, then by salts of magnesium and potassium. When evaporation reaches completion, the most soluble salts, containing potassium, crystallize, forming valuable potash deposits. For large deposits to form, complicated geological conditions must be present and must prevail over long periods, for 1,000 feet of seawater with 2.65 percent of dissolved halite will yield only a 15-foot bed of salt. For the formation of salt beds 2,000 feet thick, evaporation must continue in inland basins over long periods of time.

Salt deposits of economic importance occur from central New York along the Appalachian Mountains to West Virginia; in the lower peninsula of Michigan; in New Mexico and Texas; Alberta, Canada; Stassfurt, Germany; and elsewhere in Europe.

Iron-bearing Sediments

Sedimentary rocks exposed along highways and railroad cuts, especially in semiarid areas, often have a pronounced red color. This hue is due to the presence of the red oxide of iron, which originates in iron-rich igneous rocks. The weathering of these rocks frees the iron in a soluble or colloidal form easily transported by water. Upon reaching a suitable environment the iron is precipitated as a hydrated oxide or carbonate, sometimes with sulfur as pyrite or marcasite. During consolidation, most of the water is released and carried away, leaving the material as hematite, the principal mineral in the sediment. At times the deposition of iron becomes so heavy that beds many feet thick are formed.

Bituminous Rocks

Plant materials left on the land are soon destroyed by decomposition; the residue is lost in the surface soil or incorporated into the clayey argillaceous soils, which take on a black color because of the liberation of carbon. If, on the other hand, the plant debris is lodged beneath stagnant water, it does not decompose; instead, it accumulates, eventually forming thick deposits of peat, which are later covered by thick beds of shale and sandstone. During this long burial, the peat slowly undergoes a change, which breaks down the cellular plant structure, releasing the contained water and other plant fluids. The altered peat, under high pressure, changes first to a thoroughly disintegrated mass of black plant residues called lignite. Under further pressure and heat from overlying rocks, the lignite becomes firmly compacted to a degree known as bituminous coal —our most extensive energy resource. This coal breaks up into blocky chunks as well as fine particles. Other sedimentary rocks frequently contain bituminous materials; many coal seams contain admixtures of shale and sand.

Some of the coal-forming plants were heavy producers of spores high in combustible waxy materials. These minute particles remained afloat for a time, gradually drifting into channels of slowly moving water, where they sank and formed irregular bands of spores

SEDIMENTARY ROCKS ORIGINATING FROM CHEMICAL PRECIPITATES AND ORGANIC DEBRIS

Chemical Precipitates and Detrital Organic Materials	Name of Precipitate or Organic Material	Name of Rock
Calcium carbonate	Lime mud (marl, chalk)	Limestone
Calcium-magnesium carbonate	Lime mud (marl, chalk)	Dolomite
Calcium carbonate, sand, and organic material	Marl	Limestone
Sodium chloride	Halite	Rock salt
Hydrated magnesium sulfate	Hydrated magnesium sulfate precipitate	Gypsum
Anhydrous magnesium sulfate	Anhydrous magnesium sulfate precipitate	Anhydrite
Hydrated iron oxides	Iron muds	Hematite, limonite, goethite
Disintegrated plant materials	Peat	Lignite
		Bituminous coal
		Anthracite
Largely plant spores	Peat	Cannel coal

within the mass of peat. Upon consolidation, these spore concentrations become a highly volatile coal that produces "coal oil" and burns with a long, yellow flame like a candle; hence the name for this superior fuel, cannel coal, by corruption of the word "candle."

Structures within Sedimentary Rocks

Concretions, Septaria, and Geodes

Although there are many types of concretions, their origins seem to be restricted to one set of conditions, with variations. These are quite certainly closely associated with colloidal materials—chiefly silica, alumina, and the carbonates—and detrital sedimentary particles—sands, clays, and precipitated calcium carbonate or chalk. It is uncommon to find concretions in sandstone, except for iron-bearing nodules. An exception is the Potsdam sandstone along the St. Lawrence River, which contains layers of fine, round concretions formed entirely of sandstone.

Shale or sandy shale beds contain many horizons of flattened ellipsoidal concretions that may carry fossil remains. These concretions consist of rock materials agglomerated with colloidal material.

The recent glacial clay deposits of the northern hemisphere contain many flat, discoidal clay concretions, which were built up during the annual deposition of the clay. Occasionally a concretion survived for two or even three years before it was cut off by the rapidly settling coarse part of an annual deposition. Clay concretions occasionally possess a good bilateral plane of symmetry, which shows that the agglomerating materials were within the field of some controlling forces. It is doubtful that any significant enlargement of the concretions took place once they were covered by the subsequent annual deposition. The fact that these glacial concretions originated in unconsolidated clay indicates the possible origin of all solid shale concretions in similar early unconsolidated sediments.

The solid concretions are composed chiefly of clay. The hollow septaria and geodes, on the other hand, are more closely related to the colloidal carbonates and silica. Geodes are very variable in their mineral composition. Some are made up largely of quartz, others of quartz and numerous calcite crystals that fill most of the interior. These seem to have been formed from masses of aqueous colloidal gels. Geodes often contain a number of minerals besides quartz, calcite, or dolomite: for example, sphalerite, pyrite, chalcopyrite, fluorite, millerite, galena, hematite, and others in small amounts.

Septaria are closely related to geodes. These formations initially contained a large percentage of rock material in addition to the colloidal carbonates. As the mass of jellylike material lost its water, when the adjacent sedimentary material dried out, the outer portion formed a weak shell; as drying continued, the interior rock substance cracked and produced wedge-shaped openings on the inside walls, where calcite and other minerals crystallized. Septaria make excellent decorative specimens. In some cases, the drying process produced cracks that extend to the periphery, producing a checkered effect that gives the form its common name.

Stylolites are represented by seismiclike lines in limestones and marbles that have not been strongly recrystallized and dynamically disturbed. They also occur as short, cylindrical columns that project across a bedding plane and extend into the adjacent bed. This is produced by the pressure of overlying beds and by the horizontal contractions caused by shrinkage, which follows the loss of pore water during consolidation.

Fossil remains and the tracks of animals are common features of many sedimentary rocks and are of utmost importance to geologists and paleontologists.

Cross-bedding

Cross-bedding is a structure commonly found in sandy sediments deposited by running water. It refers to short, thin beds that seem to overlap each other, an effect caused by deposition of water that continually changed direction during formation of the rock.

Pyrite Suns

These interesting concretions are thin pyrite disks found in the coal formations at Sparta, Illinois. Nearly circular in outline, they

have diameters up to 4½ inches and a thickness varying from a thin edge at the circumference to as much as ⅜ inch at the center. Their radial structure is evident from the reflections on the poorly developed surface crystal faces. A circular ridge near the edge delineates the periphery of the first development, which was later extended by a subsequent growth that tapers quickly to the edge. These fascinating specimens developed along the bedding planes of iron-bearing carbonaceous shales, which are associated with coal deposits.

Cone-in-Cone

Cone-in-cone structures are common in many shales, particularly calcareous shales. They consist of cone-shaped portions of sediments within an enclosing conical surface. They occur along bedding planes, with the cone axis perpendicular to the bedding, and around large concretions during the process of compaction.

METAMORPHIC ROCKS

In the broadest sense, rock metamorphism consists of two phases. The first is the breaking down of preexisting rocks; the second is the construction of new ones from the disintegration and alteration products of the former. In this sense, the formation of sedimentary rocks is a metamorphic process—that is, a change of form from the previous rock to detrital substances and thence to a solid rock of many individual pieces or grains. However, we customarily think of metamorphism as a changing of the earth's consolidated rocks, both igneous and sedimentary, to rocks of entirely different appearance and often of entirely different mineral composition. As the subject has vast proportions and ramifications, we must confine ourselves to those metamorphic rocks that are derived from the common sedimentary rocks and a few of the igneous rocks.

When geologic processes buried the pre-

Conglomerate marble is composed of cemented pieces of metamorphosed limestone. This specimen is from Point of Rocks, Maryland.

viously formed sedimentary rocks deep in the earth's crust, the rocks were subjected to high heat and pressure. Under these conditions, they recrystallized, and their texture and structure changed. Such rocks are known as metamorphic rocks, "rocks that have changed."

The composition of quartzites and quartzite conglomerates is virtually unchanged from their original state as beds of loose sand grains and boulders. Their eventual formation is as solid masses of quartz, with the grains and large pieces so securely bound together and interlocked that fractures pass directly through them, rather than around as in the case of sandstone or conglomerate. While in most cases the original sand and all subsequent additional material was essentially quartz, the initial sediments frequently contain as well a percentage of argillaceous, carbonaceous, calcareous, ferruginous, and other detrital materials, which will have a strong influence on the metamorphic rock.

The limestone conglomerates—which consist of rounded, waterworn pieces of limestone or dolomite—react to metamorphic processes as limestones and dolomites. During metamorphism, these carbonate rocks are subjected to severe shearing and strong differential pressures. The processes usually produce a flow structure or a thinning out of the materials into curved lines. When subjected to high temperatures, all limestones undergo a pronounced recrystallization, which produces a coarse-grained, compact texture. Metamorphism changes the limestones and limestone conglomerates into marbles, which take a good polish and are used as decorative stones.

The complete metamorphism of slate involves the most profound changes of any of the sedimentary rocks. The first noticeable change is the greater compaction of the shale and the development of fine-grained muscovite mica; this gives the rock a distinct cleavage, which crosses the bedding planes of the original shale. The rock is now called *slate*, and has special uses for roofing and laboratory table-

COMMON METAMORPHIC ROCKS

Previous Rock	Metamorphic Rock	Changes from the Initial Rock
Sandstone and quartz conglomerate	Quartzite and quartzite conglomerate	Grains and larger pieces cemented by silica. Fractures pass through the grains and pieces.
Limestone conglomerate	Conglomerate marble	Cemented with calcium carbonate and recrystallized to marble.
Shale	Slate	Start of the recrystallization of the clay minerals. Pressure develops cleavage across the bedding plane.
Slate	Phyllite	Recrystallization continues, developing fine-grained mica. Rock becomes shiny.
Phyllite	Schist	Increased size of mica plates produces a pronounced foliation and brightness.
Schist	Gneiss	Minerals change to feldspar and quartz. Banding of light and dark minerals develops.
Limestone and dolomite	Marble and dolomitic marble	Recrystallization of calcite and dolomite. Development of accessory minerals from impurities. Rock takes a good polish.
Bituminous and lignite coals	Anthracite	Increased density and hardness and loss of most volatile materials.
Anthracite	Graphitic anthracite	Much of the carbon changed to graphite. Material will not burn as a fuel.
Peridotites, pyroxenites, basalts	Serpentine, chlorite, hornblende	Hydrothermal processes add water for conversion to serpentine and chlorite.
Hematite and iron-bearing shales	Magnetite and magnetite-rich schists and gneisses	Release of oxygen, increase in the percentage of iron minerals.

Theoretical reconstruction of the origin of the Devil's Tower, showing the supposed overlying layers that were subsequently eroded.

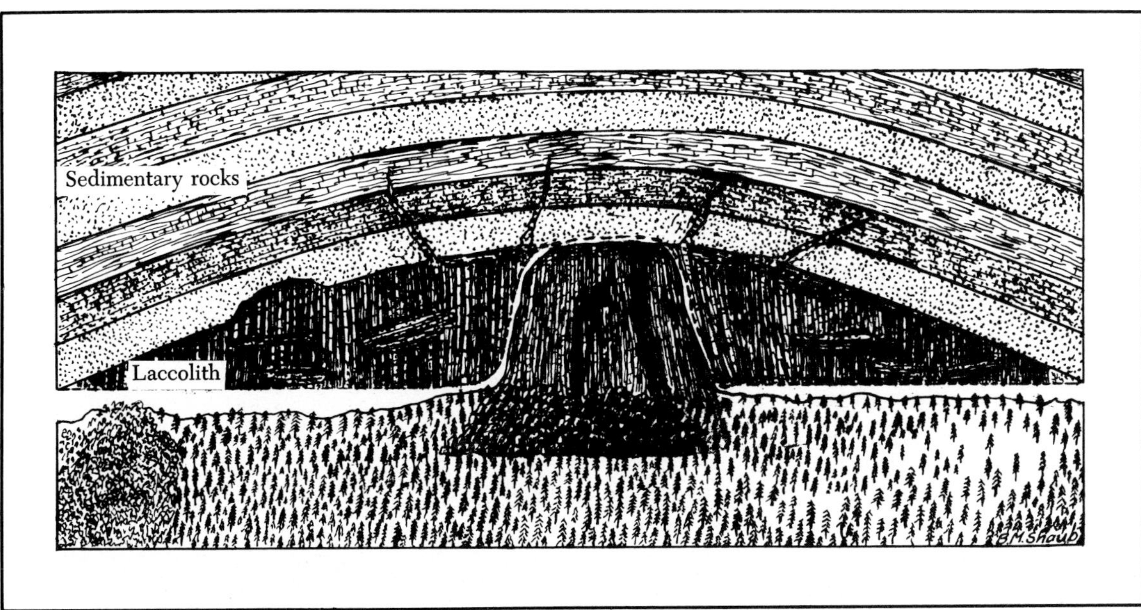

tops. Continued metamorphism develops the mica into larger grains still oriented in parallel positions. The greater development of the mica lends a shiny or glistening appearance to the cleavage plane of the rock, which is now known as *phyllite.* Cleavage in the rock does not stop at this stage.

The continuation of high pressure and elevated temperatures leads to further recrystallization and enlargement of the muscovite grains. With increased metamorphism, visible quartz grains and possibly some ferromagnesian minerals appear, provided that the composition provides the chemical elements for such minerals. The rock now has a highly glistening surface and rather coarse mica grains. Called *schist,* it splits easily along the cleavage plane of the mica grains; hence it is said to have schistocity. Accessory minerals such as garnet, staurolite, andalusite, kyanite, sillimanite, magnetite, zircon, and others appear. Continued changes develop feldspar from the mica with the release of the water of crystallization; the dark minerals aggregate into zones, producing a banded rock of feldspar, quartz, some mica, and dark minerals. The rock is now a *gneiss.* Further changes could bring about a fusion of the most easily melted minerals; if the process continued, a magma would eventually result, which on cooling would become once more an igneous rock, a granite.

When subjected to a high degree of metamorphism, iron-bearing sediments yield schists or gneisses rich in specular hematite or magnetite or both. The felsitic igneous rocks would change into schists and gneisses, while the basalts would change to chlorite schist or to serpentine rocks. The low-rank coals would change into subanthracite and anthracite, and finally to a graphitic anthracite, which is not suitable for fuel.

Contact Metamorphic Rocks

The intrusion of large masses of magma into sedimentary rocks, chiefly limestone, frequently produces a halo of metamorphic rocks about the intrusion. These rocks are composed of a wide range of minerals, depending on the amount of magmatic compounds introduced into the sediments, and the amount and kind of chemical compounds within the sediments. Contact metamorphic zones occasionally contain valuable ore deposits and frequently yield fine mineral specimens, especially if there have been enough volatile compounds or elements to hold cavities open. Minerals typical of contact zones include graphite, spinel, tremolite, diopside, and andradite.

THE DEVIL'S TOWER

Most of us look forward to trips in the field, for collecting natural objects or viewing the scenery created by complicated geologic processes. An outstanding example of an intriguing geologic phenomenon is the so-called Devil's Tower of eastern Wyoming. There can be little doubt that the first humans who saw this unique and stately structure looked upon it with great curiosity and awe; and eventually they developed numerous myths to account for its origin.

Present-day geologists have explained this origin according to several different geological theories. The most convincing one is that the tower is the remnant of a laccolith. According to this theory, the present formation represents the core and the surrounding base of the laccolith; a much larger portion of rock once existed above this base and presumably has been removed by erosion over millions of years.

It is easy to follow the geologic events that took place after the magma intruded into the rock sediments and raised the roof into a dome. The solidified magma and the conduit through the underlying rocks would have closely resembled a giant mushroom on its stem. The rocks of the area, although consolidated, are relatively soft and easily eroded by wind and water. The domed top was undoubtedly the first to be uncovered. The igneous rock was hard and tough, and remained until the erosion had reached the base and started to undermine the edges of the laccolith. As erosion continued, more extensive undercutting of the columns progressed, causing the loose, unsupported columns to collapse. This process continued until the central, resistant core was reached. Upon this central neck the remaining part of the laccolith still rests.

The mass of long, relatively thick columns in the Devil's Tower makes it a marvel of its kind, unique throughout the world. In 1906, President Theodore Roosevelt recognized its extraordinary nature, beauty, and scientific value by establishing the site as the first United States National Monument, now visited annually by many people from all over the world.

There are many other examples of columns of basalt around the world; but they are usually much shorter than those of the Devil's Tower, and most have an entirely different geologic origin. The Giant's Causeway in Northern Ireland is a most remarkable example of polygonal columns; they are so beautifully and conveniently displayed that one can walk among and over them and even find one to sit upon and rest. Germany likewise has basaltic flows in which polygonal columns have developed. In India columns were formed in the extensive Deccan basalt flows. At Mount Nonotuck along the Connecticut River in Massachusetts, the columns are uniquely exposed at their base at a site known as Titan's Piazza.

Here one may see the lower, undermined ends of the columns, which show a pronounced flaking of the basalt, caused by severe shrinkage where the columns developed next to a relatively cold layer of rock underneath. In central California there is a striking example of columns called the Devil's Postpile, still in place, along with a pile of broken columns scattered at the base of the outcrop. It is strange that the human mind should associate the image of evil with these striking and beautiful formations.

Exposed basalt columns of the Giant's Causeway extend three hundred yards along the coast of County Antrim, in Northern Ireland.

4.
Beyond Our Vision

In our search for beautiful crystals and other mineral specimens, it is rare that we have an opportunity to observe them in the process of crystallizing—except for the common mineral ice. Another exception may be seen in the arid regions of the world, where evaporation removes the water from natural mineral-bearing solutions, leaving behind white deposits of alkaline salts. Large bodies of natural brines are known as playa lakes; the much larger bodies are called salt lakes.

The weathering of pyrite-marcasite minerals in association with minerals of copper, cobalt, manganese, and magnesium produces various soluble products of the melanterite group of minerals, which may occur on rock dumps during dry periods. Occurrences like this represent only a tiny fraction of the vast tonnage of fine mineral specimens that occur within the earth's crust, beyond our vision. Most of our fine mineral specimens have been obtained through mining operations—but they represent an insignificant number of the specimens actually mined, as the great majority have gone through the crushers. Thus, the vast number of mineral collectors and the extreme scarcity of fine, undamaged crystals account for the present high prices of good specimens.

As the world population continues its runaway growth, more and more people are seeking recreation as amateur mineralogists. Hence, there is a greatly increased number of individuals involved in the search for fine minerals, and this requires greater initiative and diligence in becoming a successful collector.

Minerals are among the most ancient of the antiques that any collector could acquire. Most specimens are extremely old—not just a few thousand years, but hundreds of millions. All mineral deposits have been developed by some earth process, which usually involves physical or chemical separation and subsequent concentration. The geological processes that produce minerals are numerous and varied, and only some of them can be discussed here.

MINERAL ORIGINS

We soon learn that individual specimens have inherited characteristics that at once mark them as having originated in a specific mineral deposit. The reason for the specific character of specimens from specific localities is not thoroughly understood. Our knowledge of the deeper-seated mineral solutions and their physical and chemical properties is still largely theoretical, for it is impossible to sample magmatic solutions as they leave their sources. Furthermore, laboratory experiments cannot at present duplicate the physical and chemical nature of solutions that come from such vast and diverse bodies of fluid igneous materials.

Other mineral-forming solutions are far more accessible to mineralogists, who find that a study of minerals is greatly facilitated by classifying them according to their origin. In this way we can discover many reasons for certain mineral associations.

MINERALS FROM AQUEOUS SOLUTIONS

When water of meteoric origin—rain and snow—falls to earth, some of it starts on a long, tortuous journey downward through the rocks. Most rocks have many fracture planes and other small or large fractures or faults. Rainwater contains no dissolved solids, but as soon

Preceding pages: An iron mine near Hibbins, Minnesota. *Above:* Crystal of diamond (South Africa), found in basic igneous rocks, especially volcanic necks. *Opposite:* Striated crystals of pyrite, which occur in fissure veins resulting from hot magmatic solutions.

as it reaches the earth it begins to dissolve the rock materials with which it makes contact. This dissolution proceeds very slowly, for most rocks are very insoluble. In some areas, however, rocks that contain sulfides and other slightly soluble minerals have weathered for long periods.

In semidesert or arid regions, the water table may be more than two thousand feet below the surface; through this distance the water percolates downward, dissolving metallic minerals at higher levels and depositing them in enriched amounts at the lower ground-water level. This is the origin of many ore deposits. Many of these areas contain commercial mines built to extract the valuable ores. Tunnels have been dug to the downward-moving solutions, while air circulating through the tunnels evaporates some of the water, causing the precipitation of minerals carried by the waters. The mineral matter is deposited on the walls and floor in a wide variety of fine-grained shapes, variously described as botryoidal, reniform, mammillary, stalactitic, or stalagmitic; the layers thus formed are called crusts.

The percolating underground waters, when

passing through the joint systems of limestone rocks, usually dissolve material from the walls of the openings, and in areas where the rain is heavy, large quantities of water percolate downward for long distances. The dissolving activity of the meteoric water is increased by its acquisition of carbon dioxide (CO_2) from the air and from the ground. The addition of CO_2 yields small amounts of carbonic acid, which is an effective solvent of limestone. The water channels gradually grow larger and may join to form open spaces. The continued coalescing of smaller and larger openings eventually opens up very large spaces—rooms or amphitheaters, or caves. Caves are common along the limestone belts of the Appalachian Mountains; at Mammoth Cave, Kentucky; Carlsbad Caverns, New Mexico; and at other localities throughout the limestone areas of the world.

Once these cavities have been formed, a reverse process may set in, and the caves may gradually be filled by calcite or aragonite crystallizing on the ceiling, walls, and floor.

Most caves have enough openings to allow some movement of air, which evaporates the moisture and causes the precipitation of calcite or aragonite from the seeping water. The formations on the ceiling look like icicles and are called stalactites. (To remember the word, think of the letter C, in reference to the ceiling where they form.) Those formations that build up from water dropping to the ground and evaporating are known as stalagmites. (Again, the letter G in the word reminds us that this formation is on the ground.) The wall coverings are sometimes called draperies, for their fantastic shapes and folds. The color of these structures varies considerably; it is usually white or cream-colored, and occasionally a yellowish or dark brown, due to the presence of limonite, which is dissolved from the overlying rocks. The color variations and the configurations of cave deposits are extremely attractive and have drawn millions of visitors to these natural areas.

Heated Waters of Meteoric Origin

In areas where igneous magmas have risen to positions relatively near the surface of the earth, the surrounding rocks are often fractured, so that surface or ground waters descend to hot rocks and become heated. Since the heated water is lighter than the cold water, a circulating system can develop, in which the cold water descends and the hot water returns to the surface, as a geyser or hot spring. The hot water dissolves mineral material from the enclosing rocks and brings it upward near or to the surface, where it is precipitated when the solution temperature drops below the saturation point of the minerals. (The saturation point is the level of concentration of dissolved material beyond which precipitation will occur.) The resulting deposits are either calcareous — called calcareous tufa — or siliceous — called siliceous sinter—depending on the nature of the rocks traversed. Near the surface, the hot water channels, which follow the fis-

Lehman Caves (Baker, Nev.). Seeping water dissolved the rock, opening a large space. Precipitation of minerals then built up stalactites and stalagmites.

sures, may become coated with mineral deposits or crusts. Such deposits may be fine-grained or contain crystals in varying degrees of perfection and size, such as the large quartz crystals in the Hot Springs area of Arkansas.

Hot Solutions of Magmatic Origin

Crystallizing magmas, especially those rich in silica, may contain considerable water dissolved in the magma. Some of this water and other fluids or gases may escape into the surrounding rocks. When there are avenues of escape, such as fissures or other openings, the fluids or gases (at least in part) are able to reach the rocks near the surface, and eventually appear at the surface as hot springs. In the fissures, the minerals carried away by the escaping constituents are gradually deposited in the inverse order of their solubility. Those having the greatest solubility in the solutions are carried higher into the fissure system or are ejected with the solutions that are discharged as hot springs.

Crystallization of the contained minerals occurs with decreasing temperature and pressure along the channels, or with decreasing temperature or evaporation when discharged into the atmosphere. Magmatic solutions are usually highly concentrated and produce fissure veins and other types of mineral deposits, from great depths to the surface, depending on the continuity of the fracture system. The parts of the fissure systems that are not completely filled usually yield excellent mineral specimens, often of great beauty if not altered by subsequent processes. Fissure veins resulting from hot magmatic solutions usually carry a variety of minerals: quartz, galena, sphalerite, pyrite, chalcopyrite, cassiterite, enargite, and others. Examples of mineral deposits of this kind are to be found in Butte, Montana; Virginia City, Nevada; the mining areas of the Rocky Mountains; and in many places around the world, including the Balkans in Europe; Cornwall, England; Tsumeb, South-West Africa; and the tin mines of Bolivia.

We cannot always determine whether any isolated group of veins was produced by meteoric or magmatic waters. Veins produced by the latter usually have a greater diversity of mineral species—unless the magmatic solutions were divested of their igneous solutes at depth and acquired others from deeper sediments before forming veins near the surface.

There are many places to observe the fascinating phenomena of hot springs and geysers —in Yellowstone National Park and Thermopolis, Wyoming; scattered hot springs in California; geysers and hot springs in New Zealand; and many fine examples in Iceland.

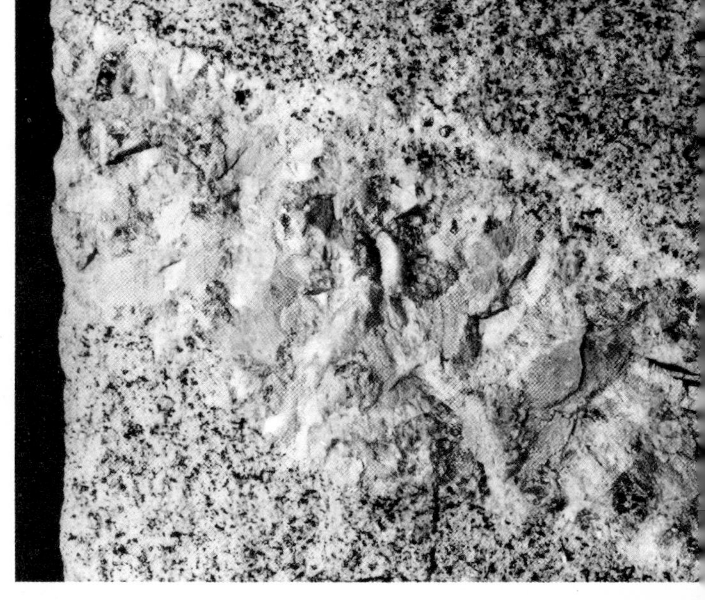

Left: Radiating crystals of wavellite (Mount Ida, Ark.), deposited in rock fissures from heated solutions. *Right:* Coarse-grained pegmatite dike, formed by injection into a granite fissure (New Hampshire).

Chemical Reaction

During igneous activity there is frequently considerable variation in the different parts of a magma, with corresponding variations in the composition of the aqueous solutions that escape. When solutions either of magmatic or meteoric origin containing certain constituents are brought together, a new and less soluble compound may form, and it may often precipitate in the region where the solutions met. Under other circumstances, solutions often dissolve solid substances from the fissure walls, and the soluble materials then in solution react to form a new substance; sooner or later this is redeposited onto the fissure walls at higher elevations. Precipitates are often of a colloidal nature and produce banded vein structures.

Evaporation

Some inland basins receive water in amounts insufficient to exceed the loss by evaporation. Such lakes consequently become rich in soluble salts. Those lakes that dry up during the arid season deposit a layer of salts on the bottom, which, over a long period of time, reaches a considerable thickness. In areas where the contributing streams gather their waters from volcanic materials, the leaching of the lavas and ash, together with the magmatic water and gases from hot springs, provides a concentration of boron salts in addition to the usual chlorides and sulfates. Such minerals as colemanite, borax, ulexite, thenardite, and glauberite occur with the more common chlorides in the salts of these playa lake deposits in the western United States, in particular in California, in Kern and Inyo counties and in Death Valley.

The water of the ocean also contains salts —halite, gypsum, carnallite, polyhalite, anhydrite, and small amounts of others. During past geological eras, vast salt deposits were laid down, extending from the Mohawk Valley near Syracuse, N.Y., to West Virginia; from Ohio to Michigan; New Mexico; Texas to Louisiana; Canada; Germany; China; and Great Britain. They likewise were produced by evaporation of vast quantities of salt water, which flowed over a shallow bar into extensive inland basins.

Crystallization in Calcareous Sediments

The carbonate sediments, which are chiefly marine precipitates, contained a high percentage of water when they first formed as a white mud on the sea floor. As new material was laid down, its weight squeezed out some of the water underneath, which then moved upward into the newly deposited precipitates. The action was progressive, and although most of the water was squeezed out, some still remained after the sediments had been deeply buried. This retained seawater is known as connate water. When under certain conditions the lime sediments were compacted to a greater extent and began to harden, shrinkage occurred with the formation of dolomite, and many cavities, both large and small, were formed. Some of these contain microscopic crystals, and others are filled with fine, large mineral species, deposited when the last portions of the saturated connate water escaped.

Magnesian limestones yield many fine mineral specimens all over the world. Some notable limestone quarries and the minerals contained in their cavities may be found near Penfield, New York—dolomite, calcite, anhydrite, clear selenite, with excellent included cubic crystals of fluorite up to an inch long; at Dundas, Ontario—the usual carbonates and many pale blue crystals of celestite; and near Clay Center, Ohio—many large-sized specimens, twinned scalenohedral calcite crystals up to ten inches long, pyrite and marcasite, rosettelike groups of dolomite, good fluorescent crystals of fluorite, and some sulfides of lead and zinc. Small crystals of other minerals are also present in limestone cavities, and the observant collector can usually add some fine microcrystals to his collection.

The dolomite wall rock surrounding the extremely rich ore body of copper, lead, and zinc at Tsumeb, South-West Africa, contains numerous cavities, lined with superbly fine crystals of numerous minerals of magmatic origin—calcite, sphalerite, chalcopyrite, galena, azurite, malachite, and many others—along with weathering products.

In Arkansas, during previous periods of hot water activity in fissures that traversed siliceous rocks, numerous crystals of quartz were deposited. Some of them are more than a foot across the sides and three to four feet long. Many excellent specimens are to be obtained from these quartz-bearing veins, among which are excellent groups of wavellite spheroids.

MINERALS FROM GASEOUS SOLUTIONS

In the vicinity of volcanic or other igneous activity, large quantities of vapors or gases are released. These consist chiefly of water vapor, carbon dioxide, and hydrochloric acid, as well as volatile compounds of fluorine, boron, and some metals, chiefly chlorides. The amounts and kinds of these gases vary from place to place. Many that are not stable react to form minerals or other gaseous products.

Mineral deposits resulting from crystallization of the various gaseous components are called *pneumatolitic* deposits. They may occur not only at the surface but at varying depths where open spaces—such as fissures or other cavities—exist. Precipitation occurs with a decrease in temperature, or with interaction between two or more gases or between the gases and solids of the wall rock of the fissures.

The minerals occuring in surface pneumatolitic deposits are chiefly sulfur, along with realgar (AsS), sal ammoniac (NH_2Cl), sodium chloride ($NaCl$), sassolite (boric acid) ($B(OH)_3$), sylvite (KCl), molysite ($FeCl_3$), cotunnite ($PbCl_2$), and hematite (Fe_2O_3). With the exception of native sulfur, few of the sublimation products that occur in the vicinity of volcanic activity are of economic importance, although some excellent mineral specimens result. Pneumatolitic deposits at great depths are characterized by the presence of a particular suite of minerals and typical altered zones along the wall rock of the fissures. The minerals most commonly associated with the process are topaz, fluorite, cassiterite, tourmaline, zinnwaldite, and scapolite and phlogopite in limestone.

The most significant pneumatolitic deposits are found in Italy at Mount Vesuvius, in Sicily, and in the Valley of Ten Thousand Smokes, near Mount Katmai in Alaska.

Sublimation: Crystallization from Vapor

In the polar zones, under favorable conditions, single crystals of ice bounded by crystal faces commonly develop. The transfer of water vapor from one mass of ice to another at slightly different temperatures usually takes place at a level considerably below freezing. The same phenomenon occurs more familiarly in containers of food in the home freezer. Frost figures on the windowpanes in cold weather also crystallize directly from water vapor in the air.

MAGMATIC SEGREGATIONS

Magmas that contain relatively large quantities of heavy minerals often separate into large masses greatly enriched in the heavier minerals. Gravity and variation in the solubility of the minerals are the chief factors governing the separation.

In the case of heavy minerals crystallizing early within a fluid magma, the crystals or grains gradually settle to the bottom of the magma, or to a level where they are supported by more viscous layers. The principal magmatic segregations of this type are chromite deposits, which occur in the Transvaal, South Africa, in the Palisade diabase, New York, and in several other localities. But the most important occurrences of segregated mineral deposits are those that develop from the accumulation and settling of sulfide minerals during the crystallization of the silicate minerals of the magma. The sulfide minerals late to crystallize form a greatly enriched phase of the magma. Material thus formed may occur in any part of the crystallized magma. Chalcopyrite, pyrite, pentlandite, pyrrhotite, and minerals of the platinum group, as well as coarse-grained pyroxenes and amphiboles, are involved in segregations. Any portion of a magma enriched in useful minerals by segregation processes may rightly be called an ore magma; it produces the so-called vein dikes that supply many massive mineral specimens. An important locality is the Sudbury Basin in Ontario.

Two things can happen to such accumulated segregations: they can crystallize in place and form more or less irregular ore bodies grading off into the surrounding igneous rock; or they can be forced into nearby fractures and thereby develop a vein system, or rich vein dikes of metallic minerals.

When injected into fissures, the ore magmas continue to separate by the same process and may produce a banded structure in the vein dikes. The central parts of these are filled with the last fraction of the ore magma to solidify. Open spaces—or vugs—in the vein dikes are frequently lined with well-developed crystals in crusts, which often provide excellent mineral specimens.

When the segregated metallic mineral solutions are injected into cool rocks, the mass of material may solidify quickly into a fine-grained mass of sulfides—galena, sphalerite, chalcopyrite, calcite—without any separation or orientation of the constituent minerals. The concentrated material may also be ejected as a sulfide mineral flow similar to a lava flow but less extensive. Samples of rich, fine-grained sulfide ores from the Buchans Mine in Newfoundland and analogous ore from the Cobalt, Ontario, area indicate an origin of close relationship to segregated ore magmas.

Pegmatites

It is quite probable that with a few special exceptions, the premier mineral deposits are the pegmatite dikes. They originate in much the same manner as the vein dikes, except that the latter are associated with intermediate and basic rocks, while the pegmatites are in general associated with granites and syenites, the most acid rocks. The size of grains and crystals in pegmatites is the coarsest of all rock types, at times reaching gigantic proportions. The shapes of pegmatites are more irregular, pinching and swelling along their length as well as vertically. They vary from stringers a few inches wide and several feet long to bodies many feet wide and a mile long. Not infrequently, large masses of the surrounding rock are included within the mass of the pegmatite. There is seldom much visible reaction between the pegmatite magma and the wall rock. Crystallization usually starts with a thin crust of pegmatite minerals perpendicular to the walls, and there is little or no orientation of crystals along the walls in the large pegmatites.

Crystallization of the acid magmas produces a residual magmatic phase rich in the most soluble constituents of the magma. Pegmatites originate from these residual solutions and occur as inclusions in the acid rocks, or, more frequently, in dikes that traverse the surrounding rocks, which are usually schists or gneisses. It is possible that some pegmatites evolved from the metamorphism of acid sedimentary rocks. Within the pegmatite magma, after it is injected into a fissure, the process of segregation through fractional crystallization continues, at times producing zones and segregations of the various minerals.

Pegmatite magmas are of two general types. The first is rich in potash and yields orthoclase, microcline, muscovite, and quartz, with few accessory minerals. Muscovite, the chief mica, is usually close to the walls; the platelike crystals are somewhat perpendicular to the walls, where the muscovite may occur in commercially attractive amounts. However, some muscovite occurs throughout the pegmatite mass. Black tourmaline is also often near the walls. Quartz is generally scattered throughout the pegmatite; however, in large deposits, it may occupy a zone near the center of the dike in large masses and may often send large quartz dikes from the central quartz core through portions of the deposit. Pegmatites have a worldwide distribution; significant occurrences of this type of pegmatite are in North Carolina, the Black Hills of South Dakota, and Oxford County, Maine.

The second type of pegmatite magma is rich in sodium, lithium, and sometimes cesium. These deposits produce cleavelandite, a bladed variety of albite, and lepidolite, a lithium-bearing mica. Spodumene, amblygonite, and pollucite as well as numerous minerals containing other alkali metals, rare earth metals, and radioactive elements may also occur. The principal micas are often lithium-bearing muscovite or lepidolite, the lavender-colored mica. Usually, neither the feldspar nor the mica are of economic importance, but they frequently make acceptable specimen material. However, there are often present rare minerals or gem varieties of some common minerals. Important sodium-lithium pegmatites may be found in Connecticut; Oxford County, Maine; South-West Africa; and elsewhere throughout the world.

In the early part of this century, pegma-

Top: Quartz crystals, formed in sedimentary rocks (Hot Springs, Ark.). *Center:* Perthite (intergrown microcline and albite), found in pegmatites. *Bottom:* Realgar, a mineral crystallized from magmatic solutions.

Top: Cleavelandite, a pegmatite mineral. *Center:* Graphic granite, a pegmatite intergrowth of quartz and microcline. *Bottom:* Grossularite garnet, which is developed in a metamorphic schist.

tites were first mined for the large sheets of mica known as isinglass, then used for lamps and stove windows. Subsequently, the deposits were worked chiefly for feldspar, the principal constituent of ceramic products. At present mica is used mainly as an insulator in electrical devices. It is also ground to a powder and used as an antifriction material or as reflecting particles for Christmas decoration. The mineral beryl—the chief source of the metal beryllium—and the metals columbium and tantalum are also products of pegmatites.

The magmas from which the pegmatites crystallized were very rich in water, sulfur, fluorine, boron, and the alkali metals. In this environment the temperature of crystallization was greatly lowered; the solution remained fluid longer, and the temperature was favorable to the long-term continuation of crystallization and to the ready migration of the chemical components through the solution. These conditions favored the growth of mineral grains or crystals to astonishingly large size.

Large beryl crystals are frequently found in quartz-rich pegmatites all over the world. In Brazil, crystals have been reported as large as nineteen feet long and five feet across. In New England, the most spectacular specimens of beryl were mined at the Bumpus Quarry in Albany, Maine. The largest of these, removed in 1950, was a tapering crystal thirty-three feet long and six feet across at the base.

Large spodumene crystals have been removed from the Etta Mine in the Black Hills of South Dakota, one of which reportedly had an estimated length of over forty feet and a weight of ninety tons.

Micas may also yield large crystals, if the magma and the associated physical-chemical conditions have been compatible. A phlogopite crystal at the Lacey Mine, Ontario, Canada, is said to have been thirty-three feet long and fourteen feet wide.

Feldspar crystals or grains may reach fifty to seventy-five feet on a side. Large crystals of quartz, topaz, and columbite have been found. The largest come from Minas Gerais, Brazil. Many other large specimens could also be cited, for pegmatites occur around the world and many of them have produced minerals unusual both in kind and size.

The following are some of the rare minerals found in pegmatites:

cleavelandite	garnet
lepidolite	rutile
beryl	biotite
amblygonite	allanite
colored tourmaline	samarskite
spodumene	autunite
triphylite	uraninite
microlite	cryolite
apatite	bertrandite
zircon	arsenopyrite
cassiterite	loellingite
pollucite	phenacite
eosphorite	gadolinite
beryllonite	herderite
topaz	pyrochlore
columbite	graftonite
tantalite	monazite
cyrtolite	petalite
lazulite	cookite
amethyst	lithiophilite
hiddenite	euxenite
euclase	perovskite
chrysoberyl	betafite
titanite	torbernite

In addition, pegmatites are characterized by various intergrowths of minerals. Of specific interest are twinned microcline and albite feldspar and a quartz-microcline intergrowth called graphic granite.

Pegmatites are the jewel factories of the granititic rocks. Further discussion of gemstones may be found in Chapter 8, but some mention of gems and their occurrence in pegmatites belongs here. Diamond, emerald, and the corundum gems ruby and sapphire are associated with basic rocks, whereas most of the other gemstones have their origins in the acid rocks, chiefly the sodium-lithium-cesium pegmatites. Most pegmatite gems are considered semiprecious stones. Good marketing techniques have given an entirely new perspective to some of the pegmatite minerals, which are excellent gemstones in their own right.

The story of the pegmatite gems had its beginning 155 years ago when two schoolboys, Elijah L. Hamlin and Ezekiel Holmes, who were interested in minerals, discovered many beautifully colored tourmaline crystals in the loose soil near the village of Paris, Maine. This

Opposite: Large beryl crystals, a pegmatite mineral (Songo Pond, Me.). *Above:* Columnar crystals of apatite (Quebec), which occur in various types of rock, including pegmatites and metamorphosed limestone.

Top: Smoky quartz on feldspar, developed in granite cavities (Moat Mountain, N. H.). *Above:* Geode from Mexico, a hollow sphere lined with lovely crystals of amethyst, a variety of quartz. *Opposite:* Blades of gypsum crystallized with sand, from Oklahoma.

locality became world-famous as Mount Mica and is at present worked sporadically for its beautiful tourmaline gems. Beyond this initial discovery, gems have been found in pegmatites around the world—especially in Maine, Connecticut, California, Brazil, India, and South-West Africa.

Pegmatites are the principal source of a number of rare elements that have important industrial applications. Among them is the element cesium, which is present in the mineral pollucite; columbium in the mineral columbite; tantalum in the minerals pyrochlore and tantalite; lithium in spodumene, lepidolite, and other minerals; beryllium in several minerals but chiefly beryl; zirconium from zircon; and the rare earth metals in small amounts from samarskite and a dozen other rare minerals. The special properties of many of these metals make them indispensable for particular uses in technical devices.

Igneous Rock Cavities

Upon cooling, all rocks contract or shrink to some degree. The resulting tension in the outer shell frequently develops open fissures; these may lead into the still fluid magma inside, which is then forced into the open spaces. Such dikes often contain large grains or crystals of some of the rarer minerals, specimens that are difficult to remove without breaking them. When the magma contains abundant gases, they often produce small cavities within the solid rock, which may contain good mineral crusts. When there has been extensive fracturing, fissures several inches or larger may be opened and crusts of minerals formed on the walls. The area around Conway, New Hampshire, is noted for the fine crystal specimens obtained from granitic veins and cavities in the granites of the area. These fissures contain fine specimens of smoky quartz, microcline feldspar, albite, clear quartz, micas, and topaz crystals several inches long.

Igneous Geodes and Amygdaloids

The gathering together of small amounts of fluid material as separate solutions within another fluid mass is known as an immiscible solution. Within a few basic rocks we find amygdules, rounded masses consisting of calcite and

zeolites or closely related minerals. Such materials appear to have separated into somewhat rounded globules during the magmatic phase. Some good specimens may be derived from the larger amygdules, from localities such as Nova Scotia, for example. In some instances the immiscible fractions aggregate into surprisingly well-formed spherical masses. When the solid material crystallizes and the fugitive materials escape, beautiful, hollow, crystal-lined spherical shells remain. These igneous geodes lined with quartz crystals, and in some instances calcite, make especially fine specimens for the collector's cabinet. Many of the finest geodes come from Mexico. Concretions of a similar origin, in Oregon and the Rocky Mountain states, are composed of an internally cracked rock material, in which the internal fractures are filled with chalcedony usually having parallel banding. They have been given the name "thunder eggs."

MINERALS DEVELOPED BY METAMORPHIC PROCESSES

Metamorphic processes continually produce changes in the rocks that are affected. Rocks, especially limestones, in contact with magmas frequently undergo marked changes in mineral composition, often resulting in extensive mineral deposits in the limestone adjacent to or overlying the igneous rocks. Mineral assemblages of this nature are known as *contact*

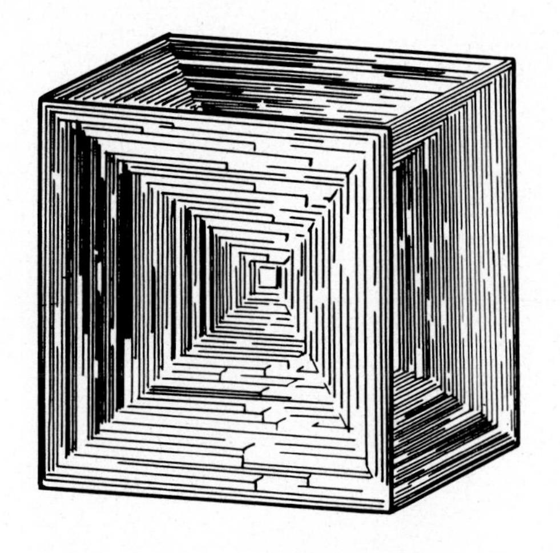

Top: Hopper-shaped halite crystal: edges have grown faster than faces. *Above:* Rosette of overlapping barite-sand crystals. *Opposite top:* Scepter quartz, an unusual combination of two crystals. *Opposite bottom:* Parallel overgrowths in a pyramidal crystal.

metamorphic deposits and are frequently the source of excellent mineral specimens. Notable contact mineral deposits are the magnetite deposits at Cornwall, Pennsylvania; chalcopyrite deposits at Clifton, Arizona; the lead-zinc mines at Magdalena, New Mexico; gold deposits at the Cable Mine, Montana; and the cassiterite deposits at Pitkaranta, Finland. The minerals found in these deposits are variable and depend largely on the composition of the invading magma. Some of the more common minerals associated with the ores are magnetite, bornite, chalcopyrite, pyrite, pyrrhotite, galena, sphalerite, gold, and silver. Other species present usually are silicates, including chiefly andradite garnet (sometimes in abundance), wollastonite, epidote, tremolite, diopside, zoisite, vesuvianite, quartz, calcite, axinite, tourmaline, fluorite, and scapolite.

Minerals of Dynamic Metamorphism

The changes due to dynamic metamorphism —involving pressure—are most noticeable in fine-grained rocks such as the shales. There are many instances when excellent mineral specimens are developed within the metamorphic rock. Among those that develop in schistose rocks are garnet, in Connecticut; staurolite, in Georgia; cordierite, in Maine; rutile, in Virginia; tourmaline; ilmenite; and magnetite, in Chester, Vermont, often in excellent small octahedrons.

The high-grade metamorphism of calcareous rocks frequently yields good cabinet specimens. The carbonaceous materials change to graphite in small masses or as small hexagonal crystals within the coarse, granular calcite, which is the end metamorphic product of lime sediments. There is usually a considerable quantity of clay and other impurities in limestones, which are changed during metamorphism into a number of characteristic minerals. Among these are phlogopite in brownish crystals; scapolite as grains or large crystals; albite or oligoclase, var. moonstone or peristerite; brown tourmaline, var. dravite; amphiboles such as tremolite and actinolite; pyroxene crystals as diopside, or as small, rounded grains scattered throughout the calcite; yellowish brown grains of chondrodite; black grains of magnetite; rutile; biotite; and wollastonite.

Large apatite crystals are not uncommon. Collectors will profit by examining the coarse-grained metamorphosed limestones for enclosed minerals.

The coarse-grained Grenville rocks of northern New York State, Ontario, and Quebec have areas of highly metamorphosed limestone. Similar rocks are to be found in many ancient-rock areas of the world.

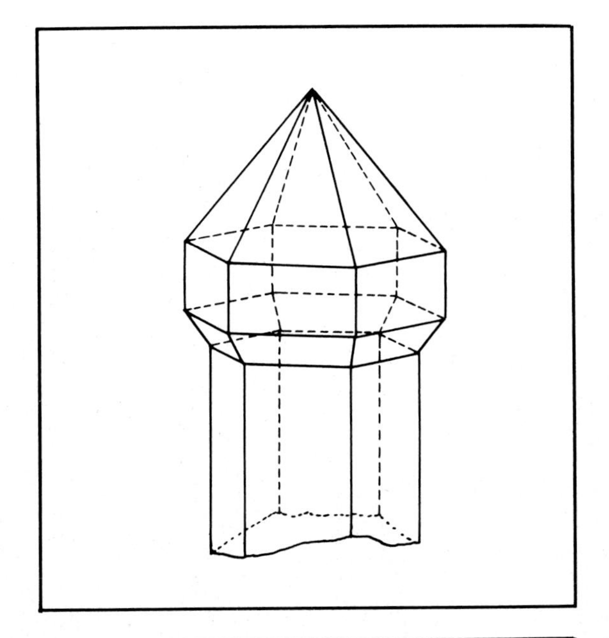

UNUSUAL MINERAL FORMATIONS

In addition to the usual formation of minerals and mineral crystals, there are some unusual situations that favor the development of peculiar crystals. One of the most interesting is the case of the scalenohedral calcite crystals of Rattlesnake Butte, South Dakota. At this locality, calcite-bearing solutions at one time coursed through the loose sand and eventually reached a saturated condition, causing the calcite to crystallize. The calcite did not act as a cement to bond the loose sand into a consolidated sandstone, but instead formed typical crystals within which the sand was enclosed. The present composition is roughly 60 to 70 percent loose sand, with the intergranular spaces made of calcite. When the crystals are removed from the wet sand, the intercrystal spaces are filled with loose sand, and the specimens appear to be only a knobby mass. But washing them under a stream of water soon reveals that they are actually a number of randomly oriented, interlocked crystals, which make unusual and interesting cabinet specimens.

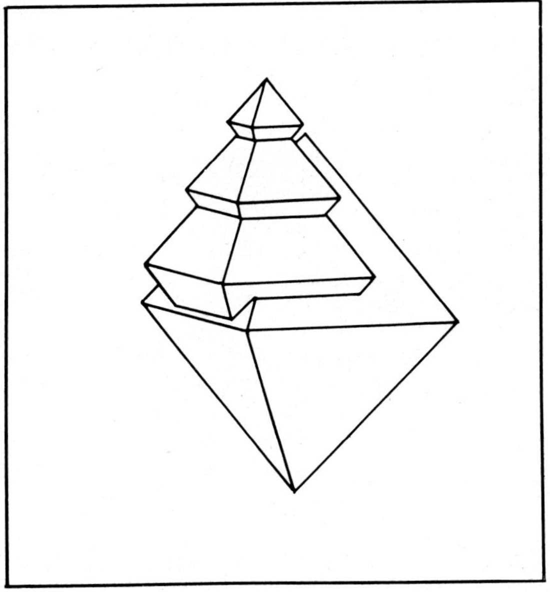

In a similar manner, solutions of barium sulfate crystallized in the red sands of Texas, and in Norman, Oklahoma, forming intriguing, beautiful rosettes of overlapping, rounded barite-sand plates. The sands that are host to the barite are finer-grained than those of the calcite-sand crystals.

The habits of crystals within the sand seem to be some of the common ones of the species. In the Great Salt Plains of Alfalfa County, Oklahoma, the solutions invading the sand carried much gypsum, which crystallized around the sand grains in thin, flat, elongated blades.

In some instances, crystals are especially active in acquiring and holding atoms along

the crystal edges or intersections between crystal faces. The result is that the edges are extended faster than the faces themselves, which remain as sunken areas. This development creates so-called hopper-shaped crystals. Halite —common salt—as it crystallizes from concentrated brines is sometimes hopper-shaped, as in the Great Salt Lake, Utah. In addition, gold crystals sometimes develop raised edges and similar sunken faces, though they are not as extreme as in the case of halite. Diamond frequently develops rounded crystals and at times shows oscillation striations on the crystal faces. Diamonds also have been known to develop hopper-shaped crystals.

Not all minerals grow by external accretion. An observer walking in the country on a cold, frosty morning past an exposed clay bank may be surprised to find the bank covered with loose, curved pinnacles of ice, which upon the slightest touch will fall away. Why? The ground and the clay are still above the freezing temperature, while the adjacent air is below freezing. Water that oozes from the clay, instead of evaporating, freezes when the air temperature is below 32° F. Once freezing starts, the additional seeping water is frozen to the base of the already formed piece of ice. The process continues until the twisted pieces of wiry ice are two to three inches long.

More suitable for collecting are the fantastic corkscrew twists of gypsum that develop from the seepage of gypsum-bearing waters in cave walls. As the solutions come in contact with dry air, they evaporate, causing the gypsum to crystallize. The crystals are deposited at the wall, and as new material is added, the gypsum form is pushed outward. Since growth at the base is not uniform and depends on seepage from the wall rocks, a twisting growth is produced. Each of these odd structures is unique, and they make a fascinating addition to a collection. However, since they are very delicate, they are rarely found except in caves that have never been previously explored, though sometimes they can be purchased from mineral dealers.

One of the real oddities among crystals is the development of a doubly terminated quartz crystal atop a quartz prism, a structure

Opposite: Sand-calcite crystals, from Rattlesnake Butte, South Dakota. *Above:* Spiral crystals of selenite, a variety of gypsum, from Texas.

known as scepter quartz. The cause of these unusual growths is not definitely known; however, quartz has two usual forms, the prism and the rhombohedron, which combine to produce the scepter. Most minerals have more than one form. For the scepter quartz, the prismatic development stops and the equant form takes over, yielding the expanded doubly terminated crystal in apparent continuity with the prismatic part. Occasionally the terminal part of the scepter is a fine amethyst, indicating a change in the solution that was the cause of the change in crystal habit. Examples of scepter quartz come from Deer Hill, Maine.

Parallel overgrowths may occur on the side of a crystal or along a crystallographic direction. The apparent repeated development of some pyramidal crystals along a crystallographic direction is an example.

The mineral witherite, barium carbonate, which is orthorhombic, always yields cyclic twins closely resembling hexagonal dipyramids. In the field the twinned specimens are short, prismatic, and tabular, with rough striated faces also in globular shapes.

Another interesting formation is the falsely formed mineral known as a *pseudomorph*—a mineral occurring in the crystal form of another species. Pseudomorphs develop when environmental conditions change following the formation of the crystal. Cubic pyrite crystals, iron sulfate, which form by metamorphic processes in slates or schists, are occasionally exposed to the air by erosion and disintegration of the enclosing rock. The sulfur of the pyrite is removed by chemical changes in the moist atmosphere, and the iron with oxygen and water becomes limonite with the same form as the pyrite cube. This is then a pseudomorph of limonite after pyrite. Many other minerals are changed into different minerals by geological processes, many of which are active within the earth's crust. Other examples are malachite in octahedrons after cuprite at Chessy, Lyons, France; galena after pyromorphite, Bernkastel, Germany; copper after aragonite, Michigan; and aphrosiderite after garnet, Michigan.

A pseudomorph of limonite after pyrite, retaining the cubic form of pyrite, from Pelican Point, Utah.

5. Tests and Tools

Determinative mineralogy means the determination of the nature, composition, and classification of minerals by means of any methods available.

In modern practice, many methods are used, sometimes requiring elaborate and expensive equipment to obtain data. All the properties of minerals may play a role in identification, depending on the resources of the laboratory or the individual making the determinative tests.

Optical tests are very powerful. The simplest involve refractive index measurements, based on the principle that when a mineral grain is immersed in a liquid of identical refractive index, the edges of the grain will seem to disappear. This occurs because with matching liquid-solid indices, light travels without bending from liquid to solid and back into the liquid; hence, no "shadow edges" are created, and the grain edges cannot be seen. Such tests are usually performed microscopically, and extensive sets of refractive index liquids, sometimes varying by as little as 0.001 unit, are available.

Additional optical tests are done with polarized light. These methods make use of the specific changes in the nature and direction of polarized light in minerals, or, for that matter, any crystalline materials. In polarized light, fabulous displays of bright colors may be seen, making this branch of determinative mineralogy fun as well as rewarding. Sometimes mineral fragments are examined, and in some cases thin slices of minerals or rocks, called thin sections, are used. A thin section must be ground thin enough to allow the transmission of light, sometimes as thin as one thousandth of an inch. Unfortunately, a good polarizing microscope is very costly, at least several hundred dollars, so complex optical tests are beyond the range of most hobbyists.

Specific gravity is a very useful test that can sometimes be definitive. Specific gravity is as much a determination of volume as of weight; with even simple balances, weights can be accurately measured. So the major problem is to measure volume accurately, because, by definition, SG is the ratio of the weight of a volume of material to that of an *equal volume* of water. A graduated cylinder or test tube, marked off in milliliters, can be used for larger grains; measurements are simple if weights are taken in grams, because the weight of one milliliter of water is almost exactly one gram. Mineralogy laboratories today employ a special balance called the Berman balance, which is extremely precise. The best the amateur mineralogist is likely to do is a specific gravity measurement accurate to about one-tenth unit. This is sufficient for all but the most difficult cases.

Luster, streak, hardness, color, associations, and crystallography can all be accurately determined by the hobbyist, with little equipment or sophistication. These tests are adequate to identify many common mineral species, but lose value in difficult cases, where several minerals have very similar properties. For these reasons the professional mineralogist today relies almost exclusively on X-ray analysis, a powerful and precise method that is capable of routine identification of the more than two thousand known mineral species. X-ray tests are quick and accurate.

Before X rays were used in mineralogy, the most powerful identification tools were the microscope and chemical analysis. A complete chemical analysis of a mineral may provide positive identification. However, several examples are known of minerals that have identical compositions but different internal atomic structures; hence they are distinct and different species. Chemical tests must therefore always be used in conjunction with crystallography and other tests.

Few hobbyists today have chemical laboratories at home, or have access to them. Nonetheless, simple chemical tests that are extremely useful in identification are within the range of the finances and abilities of the dedicated amateur mineralogist. There are two types of chemical analysis used in connection with minerals.

Microchemical analysis involves the dissolution of a mineral, to create solution species of the various chemical elements it contains. These elements are then specifically precipitated with special reagents. The precipitates are not only colored, but also have distinctive crystal forms, which are observed microscopically. The shapes and colors of the crystals

Preceding pages: Intergrowth of albite in muscovite, photographed in polarized light (N.H.). *Opposite:* Equipment used to determine mineral identity. Top, a blowpipe. Bottom left and center, Bunsen burner and the parts of its flame. Bottom right, stearine candle and its flame.

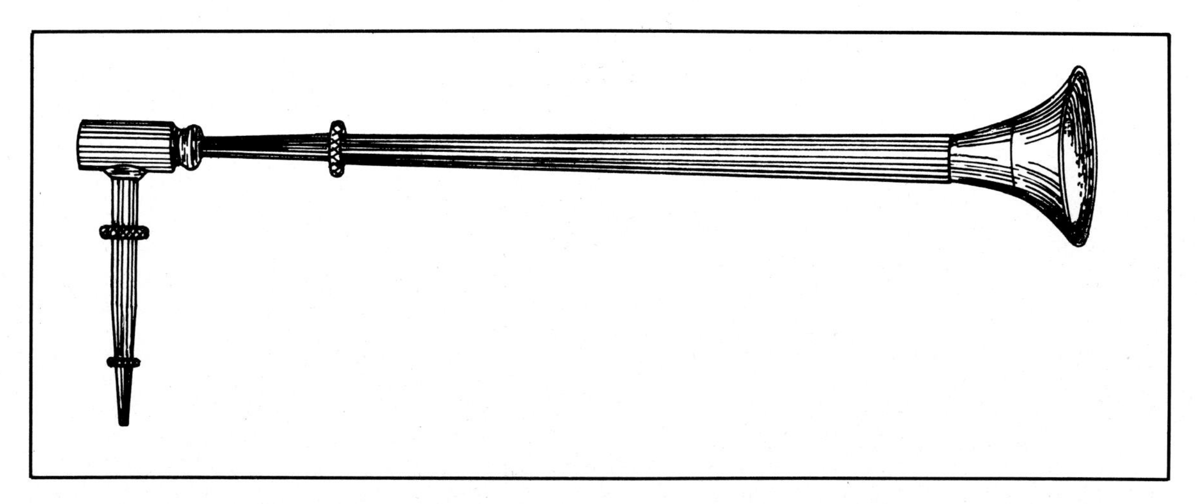

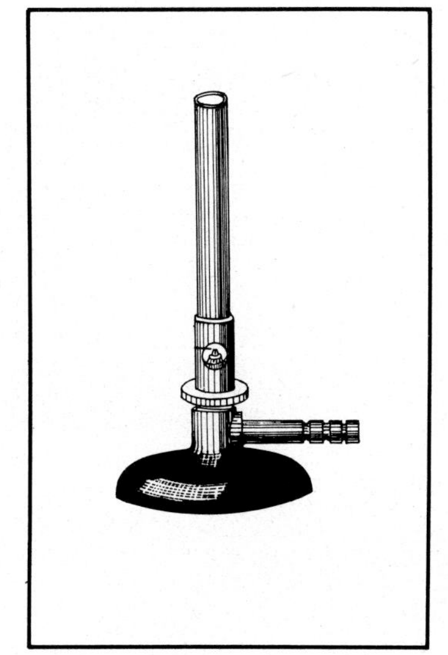

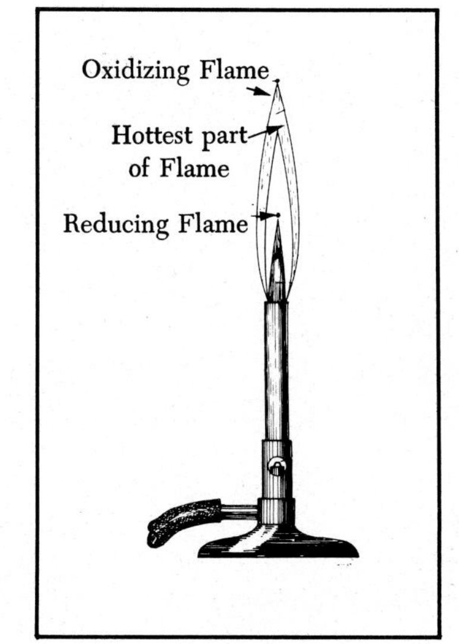

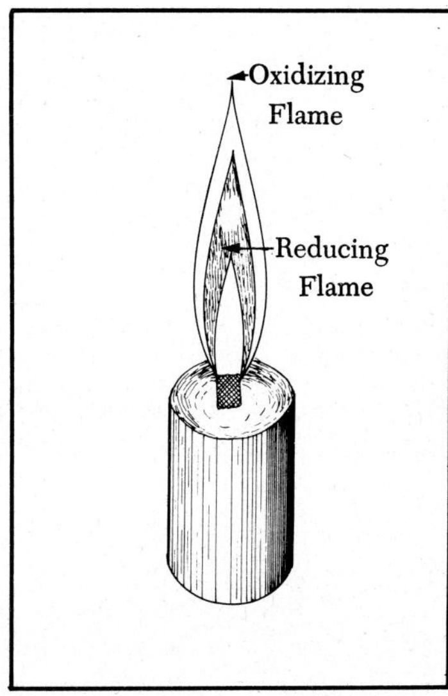

precipitated can be used to identify the chemical elements present in the original mineral.

Of far more general use is so-called blowpipe analysis. This technique has been used by mineralogists for more than a century. Even today, many college mineralogy courses include some work with the blowpipe. The outstanding advantage to the beginner is the simplicity of the technique and the small sizes of specimens that can be analyzed. The blowpipe technique is excellent for members of a mineral club, who can set up a workshop, share expenses for equipment, and consult with each other in testing minerals collected on various field trips.

DETERMINATIVE TESTS

The beginner should perform all determinative tests on the minerals he collects on trips. This will provide firsthand familiarity with the testing methods, and a knowledge of the reactions of easily identifiable minerals.

The first examination of a specimen is for crystals. If any are present, they should be studied, perhaps drawn, and measured with a simple goniometer. Knowledge of the crystal system of a specimen is a major step in identification, as it eliminates many possibilities immediately. Next, check for hardness, cleavage, color, specific gravity, luster, and streak. Charts for recording such properties, such as the following one, are useful and help in identifying the mineral.

In most cases, careful comparison of test results with tabulated physical properties will result in identification. Another major aid in this is a knowledge of the occurrence of the

MINERAL IDENTIFICATION CHART

Reference ______________________________ Date ______________

LABEL RECORD

Specimen No. __________ Collected by ______________________________
Date ______________ Locality ______________________________
Type of deposit ______________________________
Associated minerals ______________________________

PHYSICAL PROPERTIES

Hardness ____________ Specific gravity ____________ Color ____________
Luster ____________ Streak ____________ Fracture ____________
Tenacity ______________________ Diaphaneity ______________
Fluorescence ______________________ Phosphorescence ______________
Crystal system ______________________ Twinning ______________
Habit ______________________________
Cleavage ______________________________
Miscellaneous ______________________________
____

BLOWPIPE AND CHEMICAL TESTS

Fuses at __________ Flame (oxidizing) ______________ (reducing) ______________
Flame color ______________ On charcoal ______________
Sublimate ______________________ Slag ______________
Metallic button ______________ Elements in button ______________
Reaction on plaster tablet ______________________________
Open tube test ______________________________
Closed tube test ______________________________
Borax bead test ______________________________
Salt of phosphorus bead test ______________________________
Solubility in acids ______________________________
Fusion with fluxes (tests) ______________________________
____
____

Miscellaneous tests ______________________________
____
____

Distinguished from ______________________ by ______________
from ______________________ by ______________
Mineral ______________ Composition ______________
Isomorphous with ______________________________
Other minerals of the group and compositions ______________________________
____
____
____
____

specimen. As we have seen, certain minerals occur in specific types of rocks, in association with certain other minerals. A puzzling specimen may fit two descriptions fairly well, but one of these will probably be for a mineral not found in the locale from which the specimen actually originated.

If identification is not possible on the basis of physical tests alone, chemical tests can be used. These tests require some simple tools and chemicals, summarized in the following table. The amount and kinds of equipment obtained by the amateur will depend on his seriousness and thoroughness in identification. An abbreviated kit will generally suffice to start with, since materials can always be added stepwise, as the need arises. If blowpipe analysis also fails in identification, professional help should be sought at a local college or mineralogy laboratory, or at a museum or mineral shop.

Top: From Mexico, fine crystals of wulfenite, usually identified by its distinctive color and crystal forms. *Bottom:* Hemimorphite from Durango, Mexico, identified by its crystals and blowpipe reactions.

THE BLOWPIPE

The blowpipe itself is a simple but ingenious device. It consists of a tapering tube leading to a small chamber for collecting moisture. Beyond this chamber is a right-angle bend and a tapering nozzle.

The tip of the blowpipe is held just inside or just outside of the flame. You do not blow through the blowpipe. The proper technique is to inhale, close the throat as if you were swimming underwater, and puff out the cheeks. Air is pushed into the blowpipe simply by slowly collapsing the cheeks. Just before the cheeks are back to their normal position, a flip of the

THE COLLECTOR'S BASIC TESTING LABORATORY

Equipment for Blowpipe Tests

- Blowpipe
- Small propane gas cylinder and burner *or* Bunsen burner and tubes
- Hardness pencils
- Jolly balance
- Streak plate
- Porcelain spatula
- Ivory spoon
- Small agate mortar and pestle
- Metal or card protractor
- Tongs
- Small filter papers
- Test tube holder
- Test tube rack
- Diamond mortar
- Small hammer
- Platinum-tipped tweezers (optional)
- Anvil or small block of steel
- Pliers
- Alnico magnet
- Small three-cornered file
- Steel tweezers
- Scriber for testing hardness
- Platinum spoon (optional)
- Merwin screen or blue glass
- Ring stand
- Charcoal sticks 3" x 1" x 3/4"
- Plaster tablets
- Stearine candles
- Triangles for supporting crucibles
- Small crucibles
- Platinum wire .016" in diameter
- Alcohol or oil lamp
- Pin holder for platinum wire
- 10X hand lens
- Small porcelain evaporating dish
- Casserole for evaporation

Glassware

- Small test tubes
- Small beakers
- Small funnels
- Hard glass tubing 3-4mm inside diameter
- Watch glasses
- Wash bottle
- Pipette

Dry Reagents

- Sodium carbonate, $NaCO_3$
- Borax, $Na_2B_4O_7 \cdot 10H_2O$
- Phosphorus salt, $HNaNH_4PO_4 \cdot 4H_2O$
- Blue litmus paper
- Yellow tumeric paper
- Potassium bisulfate, $HKSO_4$
- Potassium pyrosulfate, $K_2S_2O_7$
- Potassium iodide, KI
- Copper oxide, CuO
- Potassium nitrate, KNO
- Granulated tin, zinc, lead
- Magnesium metal (ribbon)

Wet Reagents

- Distilled water
- Hydrochloric acid, HCl
- Nitric acid, HNO_3
- Sulfuric acid, H_2SO_4
- Ammonium hydroxide, NH_4OH
- Potassium hydroxide, KOH
- Barium hydroxide, $Ba(OH)_2$
- Calcium hydroxide, $Ca(OH)_2$
- Ammonium sulfide, $(NH_4)_2S$
- Ammonium molybdate, $(NH_4)_2MoO_4$
- Cobalt nitrate, $Co(NO_3)_2$
- Ammonium carbonate, $(NH_4)_2CO_3$
- Ammonium oxalate, $(NH_4)_2C_2O_4 \cdot 2H_2O$
- Sodium phosphate, $Na_2HPO_4 \cdot 12H_2O$
- Barium chloride, $BaCl_2 \cdot 2H_2O$
- Silver nitrate, $AgNO_3$
- Potassium ferricyanide, $K_6Fe_2(CN)_{12}$
- Potassium ferrocyanide, $K_4Fe(CN)_6 \cdot 3H_2O$
- Ammonium sulfocyanate, NH_4CNS

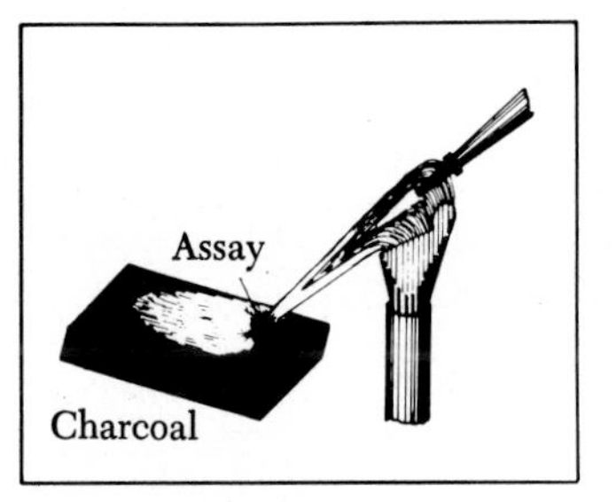

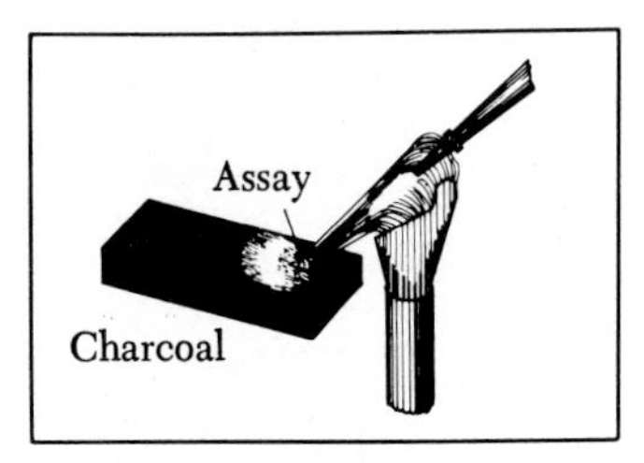

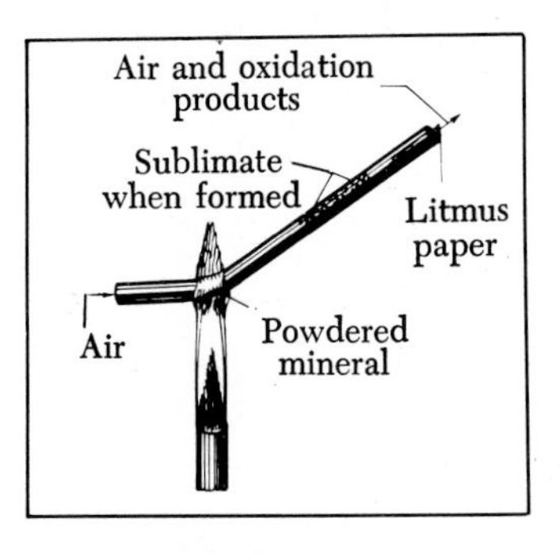

Top: Oxidation test on a charcoal block, with blowpipe flame. *Center:* Reducing specimen to a metallic globule on charcoal. *Above:* Roasting the powdered mineral in an open glass tube. *Opposite top:* Flame test for oxidation. The mineral specimen is held with forceps at the tip of the blowpipe flame. *Opposite center:* Test for reduction. The mineral fragment is held in the reducing zone of the flame.

tongue allows the cheeks to dilate again, so the flow of air into the blowpipe remains unbroken. This is simpler than it might appear. To prove this fact, close your mouth and say the word "two" with the lips pressed together. Amazingly, the cheeks fill out, even though a breath has not been taken! A bit of practice will allow you to maintain the air flow with cheeks and tongue, even while you continue to breathe normally through the nose! A proficient blowpipe analyst can maintain flow through his instrument for many minutes, without tiring or becoming winded.

Flame Sources

The Bunsen burner is a common and well-known piece of laboratory equipment. It is designed to burn both natural and manufactured gases. The operating principle of the burner consists of an adjustable air valve, which permits the entrance of oxygen (air) to mix with gas at the nozzle. This provides a suitable combustion mixture. The accessory tubes are used

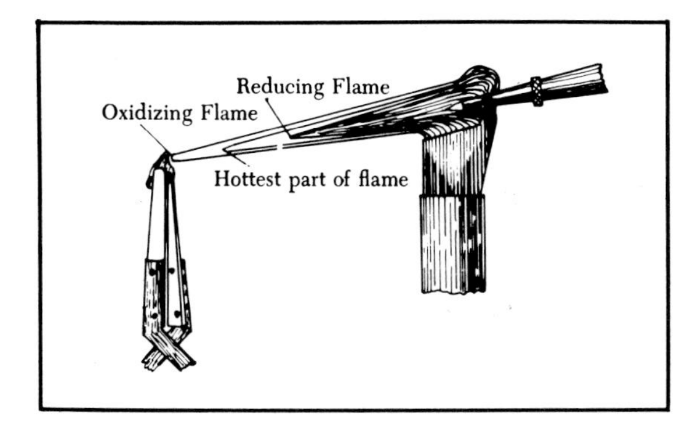

FLAME COLORATIONS

Flame Colors	Remarks
Yellowish red to orange	*Calcium* and a few other minerals when heated alone. Moistening specimens in hydrochloric acid yields more distinct colors.
Crimson	*Lithium* minerals yield distinct colored flames upon strong ignition. Carbonates and sulphates of *strontium* likewise give crimson colored flames; silicates and phosphates of strontium do not give colored flames.
Yellow	Minute amounts of *sodium* give a strong yellow flame. Whenever the yellow color is given to the flame, it may be masking other colors; hence a cobalt glass or Merwin screen should be used to permit seeing masked colors if present.
Green (yellowish)	Both *barium* and *molybdenum* yield yellowish green flames—the former in combination with carbonate and sulphate radicals but not as silicates and phosphates; the latter in the form of oxides and/or sulphides. A yellowish green flame is also produced by *boron*. After ignition, the tumeric paper test in hydrochloric acid is conclusive. *Thallium* gives a clear green color to the flame. After *copper oxide* and *iodide* are moistened with hydrochloric acid, the flame appears azure blue tinged with green. The elements *tellurium*, *antimony*, and *lead* yield pale green flames. A bluish green coloration of a pale tint is produced by *phosphorus;* even the pale color is often an aid in identification; however, *zinc* yields bright bluish color to the flame.
Blue	Sky-blue colored flames are produced by *copper chlorides* and *selenium*. The latter is accompanied by the foul odor of decayed horseradish. The blue copper chloride flame is tinged with a strong green. A pale blue color is produced by *lead*, the outer parts of which are tinged with green. A greenish blue is yielded by *phosphorus* and *antimony*, while *arsenic* yields a pale blue.
Pale violet	The alkali metals *potassium*, *rubidium*, and *cesium* yield pale violet flames. As the sources of these elements also probably contain sodium and even minute amounts of sodium mask the pale violet color, cobalt glass or a Merwin screen is needed.

inside the Bunsen burner to provide a suitable luminous flame for blowpipe work.

The Bunsen burner flame consists of three cones which are more or less concentric and grade into one another. The inner cone consists of a mixture of gas and air. It is normally colorless, but may appear light bluish gray when viewed through the flame of the outer cones. The second cone is bluish in appearance and is the cone of rapid combustion of the combustible gas. It contains incompletely burned gases, which rapidly consume the oxygen coming through the burner from the air valve. However, when mineral oxides are placed in this flame, the hot gases actually pull oxygen out of the minerals themselves. This process is known as *reduction,* the cone known as the *reducing cone,* and the flame as the *reducing flame* (RF). The outer, or third, cone is at high temperature and in contact with the air. This provides it with excess oxygen, which is readily transmitted to any oxidizable material placed in this flame. Hence this cone is known as the *oxidizing cone* and the flame as the *oxidizing flame* (OF).

The blowpipe is primarily used as a means of selecting one of these cones, and blowing it against a mineral specimen for analysis. It is clear that the blowpipe is simply a controlling device for directing either an oxidizing or reducing flame where desired.

Other flame sources besides the Bunsen burner are candles, such as those made of stearine, which burns with a yellow flame and is suitable for blowpipe work. The candle is sometimes housed in a special brass case, which is spring-loaded and keeps the flame at a uniform height. Other heat sources are alcohol lamps, oil lamps, propane in tanks, or various combustible gases in valved tanks.

The equivalent of a blowpipe flame can also be had from a small, easily handled can of propane gas with a suitable burner tip.

BLOWPIPE TESTS

The first blowpipe test is that of *fusibility*. This test is done by grasping a fragment of mineral in forceps and holding it near the hottest part of the flame. Evaluation of the results of this test is easier if you first try the test with samples of the minerals listed on the fusibility scale, to see what the results are supposed to look like.

The blowpipe flame can also be used to determine whether the specimen can be reduced; in this case the sample is held at the tip of the inner cone. Alternately, a test for oxidation is performed by heating at the tip of the outer, or oxidizing, flame.

Some mineral fragments may impart a color to the flame, which is also a useful diagnostic result.

Tests on Charcoal and Plaster

Charcoal is one of the most important substances used in blowpipe analysis. Its action on heated mineral grains is like that of the reducing flame, only more extreme. Some sulfides can be so effectively reduced on charcoal that they actually decompose, leaving a metallic globule.

Hardwood is generally used, cut into sticks about 1 x ¾ x 3 inches and charred in

SCALE OF FUSIBILITY

Number	Mineral	Approximate Temperature	Remarks
1.	Stibnite	525° C	Fuses readily in a candle flame, also in a closed tube.
2.	Chalcopyrite	800° C	Fuses in the luminous gas flame, but with difficulty in a closed tube.
3.	Almandine	1050° C	Fuses readily in a blowpipe flame, infusible in the luminous gas flame.
4.	Actinolite	1200° C	Thin edges readily fusible with the blowpipe flame.
5.	Orthoclase	1300° C	Edges of thin fragments fuse with difficulty with the blowpipe.
6.	Bronzite	1400° C	Only the sharpest splinters are rounded by the blowpipe flame.

a charcoal kiln. Other sticks of the same size are made by grinding wood charcoal to a powder and consolidating it into a stick by mixing with a binder and putting under pressure. Both types are suitable for blowpipe work, but the hardwood sticks are better, because they are lighter, more porous, and more absorbent.

In making a determination on charcoal, all traces of previous reactions should be scraped off to leave a clean surface. Mineral

SUBLIMATE FORMED ON CHARCOAL

Element	Color of Sublimate	Remarks
Antimony (Sb)	Dense white coating of antimony oxides near assay; bluish at a distance.	Less volatile than arsenic coating.
Arsenic (As)	White arsenic trioxide coating which is very volatile.	Deposited at a distance from assay, garlic odor often present.
Bismuth (Bi)	Orange yellow bismuth trioxide near assay when hot, lemon yellow when cold; white at a distance.	The coating when moistened with hydriodic acid and heated is changed to a volatile dark brown bismuth iodide.
Cadmium (Cd)	Black to reddish brown near assay; bluish black at a distance.	Very thin deposit may appear iridescent.
Copper (Cu)	In reducing flame, copper minerals are reduced to a globule of red malleable copper.	The flame is colored emerald green or azure blue.
Lead (Pb)	Lead minerals except the phosphates are reduced to metallic lead, malleable white globule.	Near the assay a dark yellow sublimate is formed; it becomes sulfur yellow when cold and has a bluish white border when the coating is very thin.
Gold (Au)	All gold minerals when treated with sodium carbonate on charcoal give malleable yellow buttons of free gold.	
Molybdenum (Mo)	Copper red coating near assay; farther from assay, pale yellow coating when hot, white when cold.	If coating is touched lightly with the reducing flame, coating becomes a fine azure blue.
Selenium (Se)	A light steel gray faint metallic coating near the assay; at a distance, white often tinged with red (selenium).	Red fumes having the foul odor of horseradish, flame color blue. The sublimate when touched with the reducing flame imparts an azure blue color to the flame.
Tellurium (Te)	Near the assay, a dense volatile coating; at some distance, gray to brownish black coating.	When the sublimate is heated with the reducing flame, the flame becomes green.
Silver (Ag)	Silver is usually accompanied by lead and/or antimony. Near the assay the sublimate is reddish to deep lilac. All silver minerals are reduced to a white metal in the reducing flame.	After prolonged heating, pure silver gives a slight brownish coating.
Thallium (Tl)	A very volatile coating at a distance from the assay, rather faint and white.	The coating when heated with the reducing flame gives a grass green color.
Tin (Sn)	Yields a faint yellow tin oxide when hot, white when cold; not volatile in the oxidizing flame.	The sublimate treated with cobalt nitrate, $Co(NO_3)_2$, and heated becomes a bluish green color.
Zinc (Zn)	Yellow coating when hot, white when cold; does not volatilize in oxidizing flame.	In reducing flame, zinc minerals are reduced to the metal, which at once volatilizes and condenses on the charcoal. When moistened with cobalt nitrate and heated strongly, the deposit becomes bright green.

powder can be blown off the charcoal fairly easily with the blowpipe. This is prevented by making a small indentation or hole in the charcoal with a knife or boring tool, in which the powder can rest. Once the top layer of powder is fused, it will not blow away.

To make a charcoal test, a small bit of mineral is placed near one end of the block. The flame of the blowpipe is then directed at the mineral, pointing down on it at about a 30-degree angle and facing *away* from the experimenter. Within a few seconds a reaction will occur, forming a cloud of smoke which deposits on the charcoal block a short distance from the specimen (where the flame doesn't reach). The residue may be a metallic globule.

Small blocks of plaster of Paris can be made by pouring a stiff plaster paste onto oiled glass and spreading evenly to a depth of about ½ inch. Before the material hardens, it is cut into blocks about 1 inch wide and 4 inches long. A smooth surface will result where the plaster has been in contact with the glass, and the blocks can be removed and stored. Sublimates of such elements as arsenic, antimony, bismuth, and mercury stand out readily against the white plaster surface.

A plaster block test is conducted in the same way as a charcoal block test, with similiar types of results.

REACTIONS ON PLASTER TABLETS

Element	Color of Sublimate	Remarks
Arsenic (As)	White over brownish black; very volatile coating.	Garlic odor may be noted due to probable small amount of arsine (AsH_3).
Arsenic (Sulfide)	Yellowish to reddish brown coating tinged with black.	Coating not very distinct and very volatile.
Bismuth (Bi)	Orange yellow when hot; color weaker when cold.	Color not very distinct.
Cadmium (Cd)	Near assay, reddish brown to greenish yellow; at a distance, brownish black.	Film is due to oxide; is permanent and is best obtained from metallic cadmium.
Carbon (C)	Brownish black, a nonvolatile coating.	Obtained from the carbonaceous materials that yield sooty deposits; example: asphalt.
Copper (Cu)		Does not form a coating.
Gold (Au)	Slightly purplish to rose-colored coating near and under the assay.	Requires very intense heat and is best seen when cold; color is not very intense.
Lead (Pb)	Dark yellow when hot.	Lighter in color when cold.
Mercury (Hg)	Dark gray, a very volatile sublimate.	For an example use merairic oxide, HgO.
Molybdenum (Mo)	Yellowish white near assay in oxidizing flame, a crystalline coating of molybdenum oxide, MoO_3.	In the reducing flame the white coating is immediately changed to a deep blue; example: ammonium molybdate, $(NH_4)_2MoO_4$.
Selenium (Se)	Cherry red to crimson in thin layer, black near the assay where coating is thick; flame is colored indigo blue.	Sublimate is due to metal, is volatile and yields reddish fumes with odor of rotten horseradish; example: metallic selenium.
Silver (Ag)	Faint yellowish coating near assay; in the reducing flame, becomes brownish and mottled.	Coating is permanent and requires high heat. Some reduced metal may also be noted; example: silver nitrate, $AgNO_3$.
Tellurium (Te)	Volatile brown to black coating, at times with edge of blue near assay.	A drop of concentrated sulfuric acid (H_2SO_4) added to brown coating and gently heated yields a pink spot of tellurium sulfate; ex metallic tellurium.
Tin (Sn)	Faint white coating.	With cobalt nitrate and gentle heat, yields a bluish green color.
Zinc (Zn)	Faint white coating.	With cobalt nitrate plus heat, sublimate becomes a grass green color.

SUBLIMATE PRODUCED ON THE WALL OF CLOSED TUBE

Sublimate	Color of Sublimate	Remarks
Water (H_2O)	Colorless, volatile liquid.	Released by all hydrated minerals. May be acid from volatile acids of fluorine, sulfur, chlorine, and other acid-forming elements.
Oxide of tellurium	Pale yellow liquids not easily evaporated, changing to colorless or white globules; solid when cold.	Derived from tellurium and some of its compounds.
Sulfur (S)	Red to deep yellow liquid, easily volatile; yellow crystalline solid when cold.	Yielded by native sulfur and a few sulfides.
Sulfides of arsenic, orpiment, and realgar	Deep red volatile liquid when hot; reddish yellow solid when cold.	From compounds of sulfur and arsenic.
Sulfantimonides	Black solid when hot; reddish brown when cold.	From compounds of antimony and sulfur.
Arsenic (As)	Black shining mirrorlike solid, close to assay.	Formed by metallic arsenic and some arsenides; when volatilized gives characteristic arsenic odor.
Mercury sulfide (HgS)	Brilliant mirrorlike solid.	Sublimate becomes red when finely powdered.
Selenium (Se)	Black fusible globules; the smallest transmit a reddish light.	A product of selenium and some selenides.
Tellurium (Te)	Black fusible globules.	Yielded by tellurium and some tellurides; may be accompanied by molten tellurous oxide (TeO_2).
Mercury (Hg)	Gray metallic liquid globule, which may be united with a long splinter.	From native mercury, cinnabar, and amalgams.
Salts of lead, antimony, arsenic, and ammonia salts	White solid.	

Metallic globules may be created with the reducing flame, or various colored sublimates with the plaster or charcoal block. The preceding tables list results of tests made on charcoal and on plaster blocks. The elements listed are a representative sample of those found in minerals; further tests for the elements are given in the appendix.

TESTS IN TUBES

Short tubes about 4 inches long are made from hard glass with an inside diameter of 3 to 4 mm. The open tube is bent at an angle of 30° to 40° about 1½ inches from one end. This provides a resting place for mineral powder inserted into the tube. The powder can thus be roasted over a flame, and the circulation of air through the tube allows oxidation to occur. A reaction will be visible in about a minute. Some elements may thus be liberated, and test

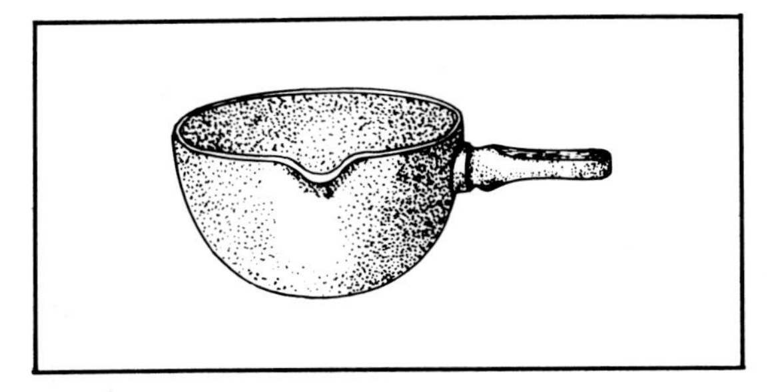

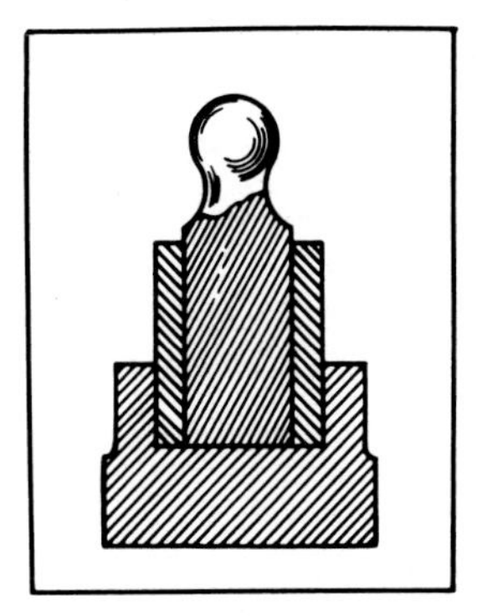

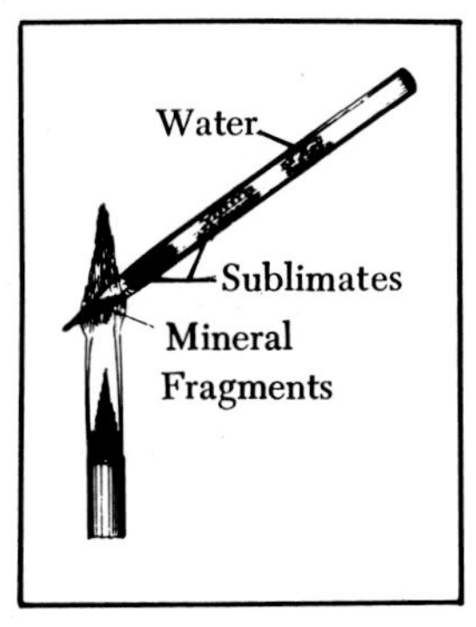

Top: Porcelain dish, for evaporating a mineral solution. *Bottom left:* Enclosed mortar, for crushing mineral fragments to a fine powder. *Bottom right:* Heating mineral fragments in the closed tube.

results noted. The closed tube is a short piece of tubing closed at one end. Mineral fragments are heated at the bottom of the tube. Heating drives off volatile substances, which may condense along the sides of the tube. Such deposits may include droplets of water, powdery substances deposited from the vapor, or "mirrors" of metals released from the fragments and carried upward in the tube in vapor form. Heating for one-half to one minute is generally sufficient to obtain a reaction.

BEAD TESTS

Platinum wire is almost essential in the chemical testing of minerals. A suitable wire consists of a 2- to 2¼-inch length of No. 26B and S gauge platinum wire, held either in a pin holder or fused into the end of a glass rod. The

BEAD TESTS WITH PLATINUM WIRE LOOP

BORAX BEAD TEST					SALT OF PHOSPHORUS BEAD TEST			
OXIDIZING FLAME		REDUCING FLAME		ELEMENT	OXIDIZING FLAME		REDUCING FLAME	
Hot	Cold	Hot	Cold		Hot	Cold	Hot	Cold
Pale yellow	Colorless to white	Pale yellow	Colorless	Antimony (Sb)	Pale yellow	Colorless	Gray	Gray
Pale yellow	Colorless to white	Gray	Gray	Bismuth (Bi)	Pale yellow	Colorless	Gray	Gray
Pale yellow	Colorless to white	Pale yellow	Colorless	Cadmium (Cd)	Pale yellow	Colorless	Pale yellow	Colorless
Yellow	Greenish yellow	Colorless	Colorless	Cerium (Ce)				
Yellow	Green	Green	Green	Chromium (Cr)	Reddish to greenish	Yellowish green	Red to pale green	Green when reduced
Blue	Blue	Blue	Blue	Cobalt (Co)	Blue	Blue	Blue	Blue
Green	Blue	Colorless to green	Brownish with much red oxide	Copper (Cu)	Dark green	Greenish blue	Brownish	Opaque red
Yellow to orange	Greenish to brown	Light green	Pale green	Iron (Fe)	Yellow to brownish	Brownish yellow	Orange to yellow	Pale violet
Pale yellow	Colorless to white	Pale yellow	Colorless	Lead (Pb)	Pale yellow	Colorless	Gray	Gray
Violet	Reddish violet	Colorless	Colorless	Manganese (Mn)	Grayish violet	Violet	Colorless	Colorless
Violet	Reddish brown	Opaque gray	Opaque gray	Nickel (Ni)	Reddish to brownish red	Yellow to reddish yellow	Reddish to brownish red	Yellow to reddish yellow
Pale yellow	Colorless to white	Grayish to yellowish	Brownish	Titanium (Ti)	Pale yellow	Colorless	Yellow	Pale violet
Pale yellow	Colorless to white	Yellow	Brownish	Tungsten (W)	Pale yellow	Colorless	Greenish blue	Greenish blue
Yellow to orange	Yellow	Pale green	Green	Uranium (U)	Yellow	Greenish to colorless	Pale green	Green
Yellow	Green	Brownish green	Yellowish green	Vanadium (V)	Yellow	Greenish yellow	Brownish green	Green

wire is used with borax, sodium carbonate, and salt of phosphorus for dissolving fragments of minerals. The wire is first heated and then plunged hot into some of the powdered chemical, which sticks to the hot wire. The mass of chemical on the end of the wire is then reheated and melted into a small globule, which is touched to a grain of mineral to be identified. The bead is reheated until the mineral fragment actually dissolves in the molten chemical, like sugar in water. When the bead cools it may attain a color due to the dissolved mineral. This color is diagnostic of the composition of the mineral, providing a clue for identification. This is the basis of the so-called bead tests.

The platinum wire is also useful in inserting mineral pastes or solutions into the flame to observe flame colors. The wire should be used by bending the end into a loop about ⅛ inch in diameter. Arsenic, phosphorus, and sulfur may alloy with the platinum, however, reducing its useful life.

OTHER TESTS AND EQUIPMENT

Forceps are required for holding specimens; the best are those that spring shut and have narrow, pointed tips, preferably made of platinum. Tongs are useful for handling crucibles and other hot objects. A ring stand with ceramic triangle is ideal for heating a crucible over a Bunsen burner. A hammer is indispensable—preferably one of hardened steel, flat at one end (for crushing) and pointed at the other (for cleaving). A mortar and pestle are handy for rapidly crushing fragments of minerals to a powder. Fragments are broken by striking a plunger, inserted in a metal sleeve that fits in a depression in a hard metal anvil. The sleeve prevents the fragments from scattering. Further crushing to a fine powder can be done in a mortar made of agate or porcelain. The former may be quite expensive, but will last for a long time.

A magnet provides rapid identification of several minerals and eliminates many others from consideration. Assays recovered from blowpipe work can also be tested for magnetism. Small Alnico horseshoe magnets are preferable.

Platinum foils and spoons are useful in fusing minor samples, but are very expensive and are easily destroyed in testing. A platinum spoon has a distinct advantage, however, in that it can be placed directly into acid for dissolving fused samples.

Glassware useful in chemical testing includes small prex test tubes, about 5 inches by ½ inch, together with a rack that holds ten to twenty tubes. Small funnels and filter paper are needed for collecting precipitates. Small Pyrex beakers are also useful in handling solutions.

Porcelain ware includes evaporating dishes and crucibles with covers. A porcelain spatula and measuring spoon are helpful in handling powdered chemicals.

Glass filters, made of colored glass, are useful in some flame tests. Some elements produce dense flame colors that obscure colors because elements in lesser concentrations are present. Blue cobalt glass filters out the red and yellow flame colorations, and allows blue or purple flames to be seen. A special filter glass known as the Merwin color screen which is relatively inexpensive, may be advantageous.

Test papers, such as red and blue litmus and tumeric paper, are useful in testing for the acidity of a solution. The color reaction of the paper is keyed to a chart provided with the paper. Many papers are available for testing for specific chemical elements.

Chemical analysis of minerals, including blowpipe analysis, is interesting, fun, and informative. It allows the beginner to feel a sense of accomplishment, because he can make substantial progress in identifying his own specimens without having constantly to turn to others for help. It is important to remember that specimens on dealers' shelves are not always properly labeled. People who attend club meetings almost always bring specimens of questionable or unknown identity. The beginner must decide how extensively he wishes to delve into the art of mineral identification. However, blowpipe analysis gives him the opportunity to dabble in whatever depth he may wish.

A complete summary of chemical tests for the elements is found in the appendix.

6.
In the Field

The search for minerals is not restricted to a few states or countries—minerals are found all over the world. And they are not unique to any particular physiographic area; they occur in all types of terrain, as well as in the seas. In the search for oil, oil drills have penetrated the earth's crust for more than five miles, and in these deep holes minerals have been found. Our knowledge of the crust of the earth extends far deeper than this, however, for volcanoes have coughed up rocks and minerals from deep within the earth, and magmas have worked their way into the upper layers of the crust. To an even greater degree, the folding, uplifting, and erosion of large segments of the earth's surface have revealed what lies beneath it.

A comprehensive study of the earth's formation and composition is properly the field of geology and beyond the scope of this book. We are concerned here mainly with minerals and rocks on or near the surface of the earth, where the average mineral collector may search for them without difficulty.

The beginning collector may wonder, Where can I find mineral specimens? The answer to this will depend largely upon where you live; for most, the best advice is to start collecting near home. Your aim should be to become knowledgeable about minerals and to assemble a fine mineral collection, rather than hoping to find large diamonds, for example, in your own backyard. Of course, there have been many instances of fabulous lucky discoveries, but the chances of equaling such finds are too remote for the beginner to foster dreams of repeating them.

SOURCES OF INFORMATION

If as a beginner you wish to learn about minerals, a good way to start is to join a local mineral club, preferably one that has an advanced collector or professional among its active membership. There are more than two thousand mineral and gem clubs in the United States and Canada, and a complete listing of them appears annually in the April issue of *Lapidary Journal.* Typical club activities include lectures, workshops, mineral and gem shows and sales, and organized field trips. The trips are the best way to find out about the mineral-bearing sites in your locale. There's a good chance, too, that you'll come home from a trip with an interesting specimen or two, which can form the foundation of your budding collection.

During the past 175 years, amateur mineralogists have made many important discoveries of mineral deposits, not only in the United States but throughout the world. As a beginner, you may feel that it's too late to enter the field, but actually there are more good minerals spots available now than there were in George Washington's day. Minerals do not enjoy the phenomenon of locomotion; they will not run away—but on the other hand, they will not jump out of the ground and surrender to your outstretched hand, either. You'll have to work hard and do a lot of digging if you want to acquire specimens. The people who have assembled the best collections are usually those most able to evaluate a mineral location and are often the best diggers.

Another good source of mineral knowledge is a privately owned mineral collection, provided the owner is agreeable to a visit. The beginner can learn much from friendly conversations with the advanced private collector, and he or she should welcome every opportunity to meet such people. Although you'll learn more readily when you can actually handle the specimens themselves—but avoid handling them more than necessary—you may also want to visit a mineral museum if you have the chance. Throughout the United States, Europe, and elsewhere in the world, governments and private institutions have assembled many thousands of mineral specimens, which they have placed on public display. Admission to these museums is usually free. In addition, several universities in the United States maintain outdoor geological gardens, in which you can find the largest-size mineral specimens. Outdoor specimens are quickly affected by the weather; many minerals easily decompose and change color, and the weathering products will eventually cover the specimens. This is a fascinating and educational feature of outdoor exhibits, which offer an excellent opportunity for study.

Further sources of knowledge are the many mineral and gem magazines in publica-

Preceding pages: The Connecticut Valley Mineral Club with a bulldozer at Smith Mine Dump, Newport, New Hampshire.

Top: Mineral collectors drilling in a railroad cut at Thomaston, Conn. *Above:* Celestite, found in limestone quarries (Dundas, Ont.). *Right center:* Cleavelandite, a feldspar found in pegmatites (Newry, Me.). *Right bottom:* Apophyllite on prehnite (diabase quarry, Centreville, Va.).

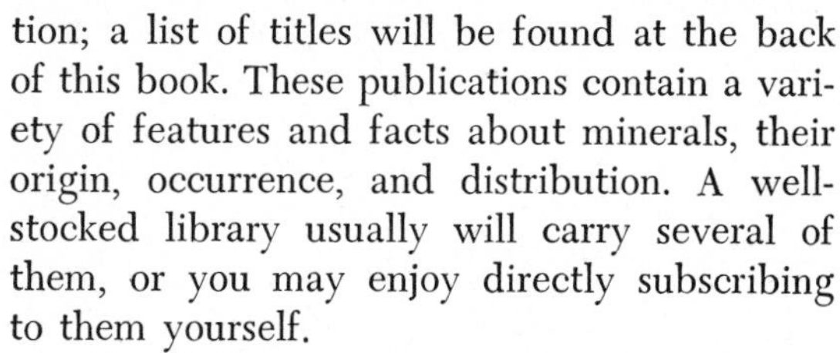

tion; a list of titles will be found at the back of this book. These publications contain a variety of features and facts about minerals, their origin, occurrence, and distribution. A well-stocked library usually will carry several of them, or you may enjoy directly subscribing to them yourself.

Finally, be sure to visit the mineral shops and mineral shows. In both places you will not only be able to see minerals from all over the world, you'll also have an opportunity to talk with other collectors. And of course you'll be able to buy the minerals you want in order to expand your collection. It is impossible to visit and collect at all mineral localities, but by purchase you can often add to your collection a specimen or two from inaccessible sources.

FIELD TRIPS

Before starting on a collecting trip, you should have a general knowledge of the area you're going to visit. If you don't, try to find someone who has been there before, and if possible, accompany him or her on a first visit. This may spare you the frustration of spending a day in a futile search for likely mineral sites that do not exist. On the other hand, once you become familiar with minerals and mineral hunting, don't overlook the fun and adventure of scouting out likely sites on your own! What are the possible places to look for? The list is surprisingly long.

Mines and Mine Dumps

The larger mining companies restrict entrance to the active work areas of their properties to all except their employees. In most instances, access to even the older dumps of waste materials is prohibited, because of the danger that visitors may injure themselves and because their presence impedes the miners' operations. Excavations in older, inactive underground mines are usually extremely dangerous, and mineral collectors should not enter them for a number of reasons. Their passages may be numerous and meandering in their course, with many offshoots, so that the explorer could easily get lost. Furthermore, the presence of people may be enough to cause an unstable roof to cave in, which could trap you and prevent your escape.

Occasionally you may be able to obtain specimens from such sources by purchase or gift. Some mining companies have made samples of their ore available to tourists and mineral hobbyists, at a convenient place away from the main work area. Some mining communities have even established museums where the products of their mines are displayed—an attractive method of advertising and a delight to collectors. These museums frequently have specimens of local minerals for sale, and the curators can often provide you with valuable advice regarding collecting sites in the vicinity.

The dumps of some old abandoned mines are excellent collecting sites. Although in most cases the valuable ores have been carefully channeled to the mine mill, the vast dumps of waste rock frequently contain pieces of branching veins or small pods of minerals that make

worthwhile specimens. The nonore minerals unearthed in the mine are normally discarded to the dumps, where industrious and knowledgeable collectors can find rare minerals of much scientific interest, and occasionally even a new mineral. Where mine dumps are being trucked away for landfill or for use in road construction, collectors should keep a careful watch for excellent specimens.

Prospect Sites

These sites always offer the possibility of yielding mineral specimens, for it was the discovery of minerals at the place that first induced the prospector to make an excavation. He was probably searching for gold and silver, but when the prospect did not pan out, he abandoned it. The loose material may not prove profitable for today's collector, either; however, it is worth a careful look around to see what the waste material contains. There are many thousands of such sites in mineralized localities around the world. Could they all be worthless to a sharp-eyed mineral collector? During the early mining ventures in Arizona, prospectors would frequently work on a small showing of altered copper minerals, sending to the dumps thousands of fine, small mineral specimens before abandoning their prospects as noncommercial. Many of these dumps may still contain fine copper, vanadium, and molybdenum mineral specimens.

Rock Quarries and Dumps

The most active quarries at present are those being worked for road materials and for building, decorative, and monumental stone. Such quarries chiefly furnish such materials as granite, limestone, marble, sandstone, and diabase. Because of acts of vandalism inflicted by trespassers, most active quarries are off limits to mineral collectors. But it is possible for the chairperson of a mineral club to arrange for a group visit on weekends, when the quarries are inactive. If you are admitted, it is wise, of course, to adhere strictly to the rules and instructions of the management.

Of the types of quarries mentioned above, the marble and sandstone quarries are the least likely to provide an array of good specimens. Cavities in limestones, however, especially dolomitic varieties, are often the source of many fine minerals.

One of the most prolific types of quarries is that developed in the working of pegmatite deposits. Pegmatites—coarse-grained granite dikes—are among the most abundant storehouses for minerals, both common and rare species. As many as fifty to seventy species have been found in some of the sodium, lithium, and cesium pegmatites, along with crystals of feldspars, quartz, and spodumene of immense size. When feldspar and quartz for ceramic uses were obtained from the pegmatites, the quarries were continually supplying the dumps with fresh discards, which frequently contained many fine specimens, often of the rarer minerals. Many of the large dumps still contain specimens, but hard digging is usually necessary to bring them to the

Opposite left: Hexagonite, found at quarries in Talcville, N.Y.
Opposite right: Calcite from a fissure cavity near Shelburne, Mass.
Above left: Heulandite crystals on diabase, from Poona, India. *Above right:* Silver-safflorite dendrites from vein systems at Cobalt, Ont.

surface. When these dumps are used for fill, the nearby collector should keep on the alert, frequently checking over the new exposures as well as the places where the material is used.

Quarries in the diabase formations are often rich in zeolites and other minerals. Such quarries are located at many places: in New England, along the diabase sheets of the Connecticut Valley; in the diabase sheets of northern New Jersey; at Centreville, Virginia; and along the cliffs of Cape Blomidon, Partridge Island, and other Bay of Fundy localities in Nova Scotia. The basaltic rocks of India have produced many fine zeolite specimens.

Mineral collecting has been prohibited at most quarries, which have become large operations, because of the extreme safety hazards involved in allowing visitors to climb at will among the rocks. It is true that some minerals are detrimental to a small extent to the quality of the crushed rock, but the probability of injury is too great for the quarrying companies to allow collectors access to the broken rock—even though collectors may be of use in picking out unwanted materials from their product. A collector's best hopes are to purchase specimens from workmen who may have picked up a few of the pieces.

There are many other useful rocks and minerals to be recovered by quarrying and by small, underground mining operations. The talc and asbestos deposits of New York, Vermont, and the southern Appalachian states have provided some fine specimens of the associated minerals; the best of these find their way to local mineral stores or into the hands of individual miners, who are usually willing to make them available to collectors. Talc, soapstone, and serpentine quarries and mines are usually the result of hydrothermal metamorphic changes. Accompanying the development of talc and the other economic materials, there is usually present a suite of associated minerals that provides the collector with some excellent specimens.

When searching for specimens in quarries and mines, the beginner would do best to work with experienced individuals. The present high price of good specimens may encourage more companies to arrange for the collection and sale of good specimens from their mines, since

Top: Pyrope garnets (Navajo Reservation, Arizona). *Center:* Chabazite from the beaches of the Bay of Fundy, Nova Scotia. *Above:* Water-tumbled agates, from gravel deposits on the shores of Lake Superior.

an especially good specimen may be worth several truckloads of the average ore.

Roads, Railroad Cuts, and Building Excavations

The grandiose expansion of railroads and highways in the past hundred years, a worldwide endeavor, has exposed many fine mineral sites. If you live near a highway presently under construction, check the new road cuts frequently and in detail on weekends. Do not overlook the small cavities, for these frequently contain magnificent microcrystals. Along the course of a new highway many different kinds of rocks are cut through, and each kind could contribute a different suite of minerals that may make good additions to your collection.

Recently, a group of geology students and instructors were examining the rocks in a large highway cut across the Conway schist in western Massachusetts, near Shelburne, when a student discovered a small group of white crystals in a small opening. This led to an expanded fissure cavity that proved to be lined with hundreds of fine calcite crystals. The calcite crystals gave the collectors the tremendous feeling of satisfaction and the thrill of discovery that is continually attracting people to the hobby of mineral collecting.

Excavations for buildings, tunnels, and subways have often removed rocks enclosing mineral-bearing cavities, from which workmen sometimes liberate the specimens before they are discarded on an inaccessible dump. Areas of metamorphic rock, especially those that have interbedded layers of calcite, are likely to provide excellent mineral specimens. Collectors will do well to check any local construction work that involves rock excavation for foundations.

Caves

There is a fascination about exploring limestone caves, but the excitement can hardly be said to stem from any spectacular minerals that may be collected. Caves are frequently lined with travertine and stalactites, but very few of these make suitable cabinet specimens. Minerals other than calcite, however, are found in limestone caves—aragonite, selenite, and niter. In weathered deposits, the original minerals have often been broken down by the dissolving action of water seeping through the rock, which results in a considerable exchange of mineral components. When the resulting solutions are carried to lower levels in the rock, open spaces are often left behind; these may contain a wealth of excellent mineral specimens—malachite, azurite, quartz, and selenite, to name just a few. One of the rarest and most unusual cave formations are the so-called cave pearls, formed in depressions in the floor into which water drops from the ceiling. Turbulence and evaporation of the water develop rounded, pebblelike concretions.

Rock Cavities

Different kinds of rocks may contain cavities, openings that are much smaller than the caves found in limestones. Such cavities originate in various ways. In granites and other crystalline rocks, they are usually formed in the final stages of cooling of the molten magma. Shrinkage upon cooling allows spaces to form, which may rapidly fill with residual water expelled by the magma. This water contains many dissolved components, which may ultimately deposit crystals in the open space filled by the solution.

In metamorphic rocks, especially strongly folded rocks, cavities sometimes develop on the concave sides of sharp folds. These are areas of tension, rather than compression, and such open spaces may fill with mineralized solutions that can deposit crystals.

In crystalline rocks, the cavities often yield superb crystals of quartz of several showy varieties. These crystal cavities also produce groups of feldspar crystals that are often studded with clear or smoky quartz crystals. These specimens may also feature crystals of some of the rarer minerals, such as topaz. Such a locality is found near Conway, in eastern New Hampshire.

The charming species of the zeolite group usually occur in cavities in basaltic rocks known as trap rocks. Cavities up to a foot in diameter sometimes produce exceptionally fine crystals, often surprisingly large for the size of the cavity containing them.

With the passage of time, rocks may weather sufficiently for crystals to be loosened and fall to the bottom of a cavity. Further weathering may expose the cavity and allow the formation of sufficient soil to hide the crystals completely. The soil accumulates in the cavity to the point where a small bush may grow in it—in fact, the presence of small bushes on a rock outcropping may well indicate that cavities are present. A metal rod four or five feet long is effective in probing the soil beneath such bushes. If you find any crystals, you can remove them easily in most cases, but take care afterward to restore the site to its original condition.

An excellent example of crystals formed in cavities—one known throughout the world—is the metamorphosed dolomitic limestone in the southern foothills of the Adirondack Mountains, in northern New York State. This formation is known as the Little Falls dolomite. During its metamorphism many small cavities formed, varying greatly in size. Their contents consist almost entirely of brilliant, perfectly formed, doubly terminated quartz crystals, ranging in size from a fraction of a millimeter to more than 20 centimeters. A cavity may contain a single crystal two to three inches long, or more than a thousand very minute and perfect crystals about a millimeter in diameter. In some localities along the formation, which extends for many miles, single crystals or clusters of less perfect but much larger quartz crystals may be found in eroded and soil-filled cavities near the surface. The crystals from this region of New York have been called "Herkimer diamonds," after Herkimer County, New York, in which a large portion of the dolomite formation is located.

Fissure Veins

All kinds of rocks are subject to fracturing. This occurs as a result of various processes. Shrinkage, or tension, can create open spaces by pulling rocks apart. Compressional forces, such as those caused by earth movements or the pressure of an igneous intrusion, can fracture rocks by placing them under severe pressure or shearing conditions that exceed their breaking strength. Such fractures may later fill with solutions containing dissolved minerals. The solutions may have been generated by nearby cooling igneous bodies, or may have

traveled a considerable distance through the surrounding rocks. The deposition of minerals in a fracture by such solutions results in what is known as a fissure vein.

A region may be subjected to different types of forces at different times. A period of compression during metamorphism, for example, might be followed by tensional forces when the pressures are released. This can create various sets of fractures in the same set of rocks. The different fracture sets may be mineralized by solutions at very different times, developing fissure veins with totally different mineralogies within the same geographical area.

Some notable vein systems and the minerals they have yielded are at Butte, Montana —a variety of copper minerals; the Sierra Nevada, California—native gold and gold tellurides; Cobalt, Ontario—silver and cobalt minerals; and Cornwall, England—copper, lead, zinc, tin, and uranium minerals, highly valued by collectors and museums. The vein systems and cavities of the Alps and the Balkans in Europe are also famous for a variety of fine mineral specimens.

When cavities remain open in the center

of the vein, they afford a great variety of beautiful crystals, some of which may be of gem quality. No partially filled fissure is too large or too small to yield splendid minerals.

Most areas of fissure veins are off limits to the amateur collector; however, many rock shops carry pieces from famous old localities. Such specimens add greatly to a cabinet and to the satisfaction of the owner.

Animal Burrows

In areas where coarse-grained rocks have disintegrated, large grains of the resistant minerals frequently accumulate. Granitic rocks seldom contain fine crystals; however, some of the basic rocks contain large, clear grains of olivine (peridot) and deep red almandine garnets, which are frequently large enough to provide excellent gems. The weathered-out resistant grains accumulate as placer deposits in gullies and streams, where they settle to the bottom and may remain buried in the ground even after the stream has dried up or changed its course. In digging their nests, small animals, such as rodents and even ants, carry the colorful stones to the surface, where they may often be picked up in numbers. On the Navajo Reservation in Arizona, for instance, gem garnets and peridots are often brought to the surface in this way.

Cliffs, Bluffs, and Rock Outcrops

The waves of the ocean are continually eroding the shoreline, working landward against cliffs and bluffs. When these formations contain mineralized cavities or fissure veins, the wave action exposes the minerals as sections of the cliffs fall to the beaches. Most of the falls occur when the frozen cliffs thaw in the spring. At this time, after it is safe to enter the area beneath the cliffs, you may find many fine mineral specimens. The beaches on the shores of the Bay of Fundy, Nova Scotia, have furnished much material from the neighboring high basalt cliffs, whose cavities carry fine zeolite crystals, quartz, amethyst, and other minerals.

Rock outcrops that are especially rich in muscovite or lepidolite mica can be useful guides to important sources of minerals. An examination of such outcroppings may reveal, for instance, the contact of two different kinds of rocks, such as igneous and sedimentary. Ig-

Opposite: Copper float deposited in glacial debris and found near Calumet, Michigan. *Above:* Consolidated clay concretions, found in some of the glacial clays in Northampton, Massachusetts.

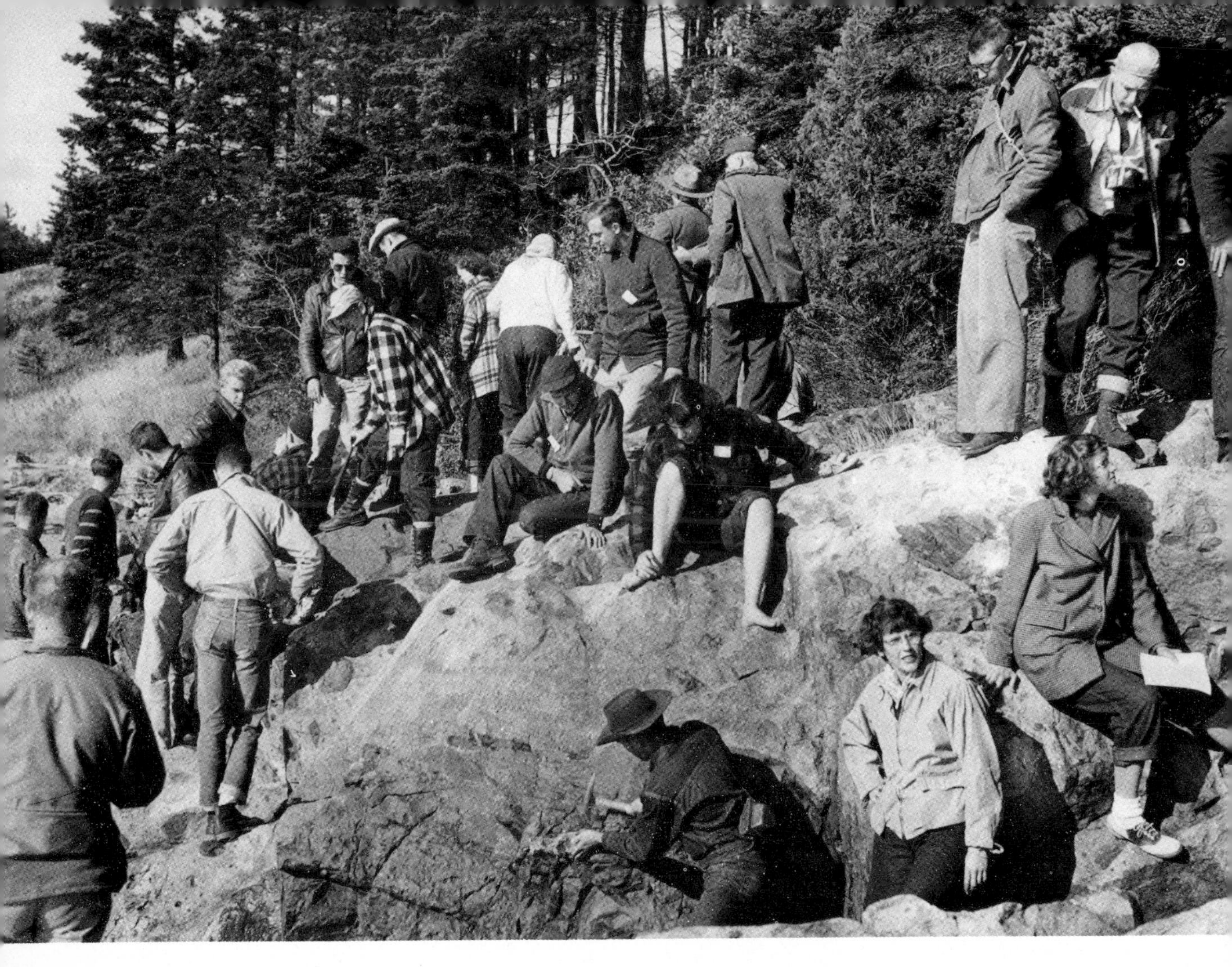

neous contacts along the sediments are often favorable places for mineral deposits and for unusual silicate minerals. The collector is frequently rewarded by examining granite outcrops, if there are pegmatitic phases present. Both small and large cavities are sometimes associated with the coarse-grained zones and may contain excellent crystals of quartz, tourmaline, feldspar, mica, garnet, topaz, and other minerals.

Deserts

It is commonly believed that deserts are mere sandy wastes, as devoid of minerals as of organic growth, but this is far from true. Collectors in the western part of the United States find many minerals in semiarid places from which they cut semiprecious stones—agates, various kinds of multicolored jasper, garnets, peridots, chalcedony, and other varieties of quartz and copper minerals. One of the prizes is the precious opal found in quantity in the arid areas of Mexico and Australia. The semidesert regions of Chile are rich in copper and nitrate minerals.

Glacial Gravels

Scattered over large parts of the northern hemisphere are vast areas of gravels, which were deposited by the retreating ice sheets of the last glacial period. These ice sheets literally swept the ancient soils and broke down material of the existing rocks. The materials carried by the glaciers included the most resistant substances released by weathering. Therefore, we frequently find remnants of rocks and their mineral contents at sites far beyond their sources. The original mineral deposits can be located by following the trail of mineralized boulders back to their origin. Among the most resistant materials are the siliceous minerals—chalcedony, jasper, and agate nodules. The glaciers that swept across the basalt area of Lake Superior carried to the south enormous numbers of agate nodules that were freed from the basalt. In parts of Minnesota and Wisconsin, some of the gravel was left in stream, lake, and other gravel deposits. Collectors in these areas search the gravels for the fine, colorful Lake Superior agates, some of which are as large as three to four inches in diameter.

The metal detector is a marvelous device to use in searching for metallic minerals that have been left in glacial debris or in the soil. A remarkable discovery of this sort was made in 1971 by a young boy near Calumet, Michigan, when he discovered on a neighbor's farm a piece of native copper weighing 9,392 pounds. The piece is now on exhibit in an outdoor museum in the city of Calumet.

Stream Beds and Beaches

Stream beds and beaches are areas where resistant minerals accumulate after being freed from the rocks in which they were formed. Some of these old riverbeds and beaches are now integral parts of consolidated sedimentary rocks. In either case, such placer deposits, as they are called, are excellent places to prospect for minerals. In the Yogo Gulch in Montana, sapphires may be recovered from a stream and its flood plain. Burma, Thailand, and Ceylon are famous for the rubies, sapphires, and other gemstones recovered from gem gravels. The state of Minas Gerais, Brazil, has yielded diamonds and topaz from streams, flood plains, and old land areas. The stream beds of the Orange River, its drainage area and flood plains, and finally the beaches of West Africa have returned many fortunes in diamonds and other gemstones. The stream beds of California rivers, flowing from the Sierra Nevada gold-bearing veins, have yielded millions of dollars worth of gold. This gold belt extends northward through British Columbia and into Alaska.

As an experiment in exploration, it could be well worthwhile to dig for some distance below the surface of a beach or riverine sandbar in some mineralized area and to pan the bottom sands. They might prove to contain interesting or even valuable minerals—monazite, zircon, ilmenite, corundum, ruby, sapphire, magnetite, and very occasionally gold or diamonds. You might enjoy putting yourself for a while in the role of a legendary prospector of some 150 years ago. Collectors who have panned the gold-bearing streams of western Maine, central Vermont, Virginia, and Georgia—or similar localities in other countries—can proudly exhibit a small vial of gold from their panning operations. It may not provide a substantial income, but the reward, in other ways, is sure to equal the effort!

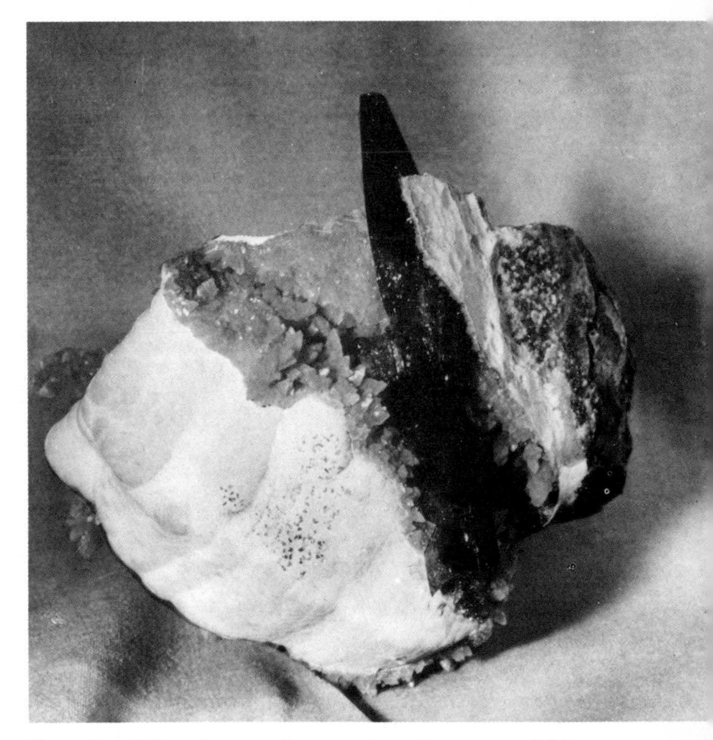

Opposite: Mineralogy students and teachers on a field trip, Mt. Desert Island, Maine. *Above:* Septarium from Wasta, South Dakota, containing a dark brown crystal of barite (center) on a bed of paler calcite crystals.

INTERESTING MINERAL FORMATIONS

Concretions and Geodes

A concretion is a consolidated mass of material, usually round; its composition is similar to that of the rock that encloses it, but it contains in addition a binding medium, consisting chiefly of calcium carbonate or silica. Shale concretions are usually made of clay bound together with calcium carbonate; they are called clay stones when they occur in unconsolidated—unhardened—clay. Concretions of various flattened shapes often possess a surprisingly bilateral symmetry, and are found typically in the Champlain clays of New England and elsewhere in the glaciated areas of the northern hemisphere. When concretions occur in sandstone, they are cemented with silica. Examples of these, surprisingly spherical in outline, occur in the Potsdam sandstone of northeastern New York State and in adjoining Canada along the St. Lawrence River.

The chalk beds of England and France contain numerous chert or flint concretions, many of which are extremely irregular in shape. These siliceous bodies originated in the chalk at the time the lime carbonate was deposited. Under the physical and chemical conditions present during deposition, the silica and carbonate remained separate. One formed fine-grained calcite, the other agglomerated into ellipsoidal and irregular knobby forms, such as silica gel, which later hardened into flint. These concretions are dark gray to black and have very little attraction for mineral collectors. They were formerly used in mills for fine-grinding crushed ores.

Concretionary forms from the Pierre shale of the Dakotas range up to three to four feet in diameter and often contain fine mineral specimens. In some instances the concretions were formed about the shells of ammonites or other creatures of the sea where the argillaceous muds were laid down. Others developed into the form known as a septarium. In the western area of the Dakotas, a pleasant collecting outing may be had by following along a stream bed to locate septaria that have weathered out of the black rock. As you walk along you will find the remains of broken septaria, evidence of previous visits by collectors. The area may have been carefully searched before, but finally you will find a concretion that has been overlooked, or given up as worthless or too large and solid for the light hammer. Not all concretions yield fossils or choice mineral specimens. However, you might be fortunate enough to locate some fair-sized brown barite crystals (which are brittle and rather easily broken) accompanied by yellow calcite, which alone can make an acceptable cabinet specimen.

A geode is a hollow nodule lined with crystals. When the interior consists of concentrically banded silica, it is called an agate geode. When the silica filling is not banded, the nodule is called simply a mass of chalcedony. Geodes are of either sedimentary or igneous origin and usually vary widely in their mineral content.

Many of the sedimentary geodes have weathered from the parent rock and are to be found loosely scattered in the soil; or, if they have been carried by water to streams, you may be able to find some there that have been overlooked by previous collectors. In a few areas they are quarried for the mineral-shop market. The presence of the rarer metallic minerals in geodes increases the desirability of the specimens.

Geodes are found in many areas of the world. In the United States, the region around Dubuque, Iowa, with its excellent crystal-lined geodes derived from sedimentary rocks, is the best known. In the 1930s and 1940s, collectors in Oregon gathered nodules filled with bluish chalcedony, sometimes banded, with an agate or irregular structure. They were unusual in that the shell had shrunk during formation, producing a rough and irregularly star-shaped section when the specimen was cut across. These nodules were collected and marketed under the name "thunder eggs."

Chalcedony and Jasper

Chalcedonies and jaspers are inexpensive decorative stones. They are usually the result of sedimentary processes, often of limited extent, and occur as concretions or small lenslike beds. Chalcedony is a variety of silica. The pure material is almost colorless, usually translucent, and of a light bluish hue. Seldom pure, it often occurs in areas containing manganese and iron-bearing solutions. Precipitates of manganese and iron oxides mixed with silica create colorful mosslike inclusions in the hardened chalcedony, which resemble organic growths. Iron oxides also provide colorful and intricate inclusions. These precipitates have led to the name moss agate. Chalcedony of uniform color (usually by iron), irregularly banded or flecked with various inclusions, may be called jasper. A characteristic of jasper is its variety of colors and interesting patterns, making it extremely desirable to people who enjoy cutting and polishing gemstones.

Altered Minerals

Secondary or alteration products are good clues to prospectors searching for ore deposits. Certain minerals are changed into secondary products when exposed to weathering—sulfides, antimonides, arsenides, some oxides, carbonates, silicates, and other mineral compounds. The primary copper minerals are prone to alter to several colorful alteration products. The metal may change to cuprite and then to azurite or malachite. Cuprite, a copper mineral, changes to the colorful green malachite. The octahedral crystals of cuprite from Chessy, Lyons, in France, make excellent small cabinet specimens; they are coated with bright green malachite and are eagerly sought.

Among the less colorful minerals, stibnite changes under weathering to a light brownish mineral called stibiconite, while ilmenite alters to the dull earthy mineral leucoxene. The pink manganese minerals, rhodonite and rhodochrosite, alter to the black mineral wad. Among the colorful alteration minerals are autunite, torbernite, gummite, curite, and other uranium minerals resulting from the alteration of uraninite. Iron minerals revert to the brown earthy material called limonite. Nickel and cobalt minerals provide the colorful alteration products annabergite and erythrite, respectively.

COLLECTING SPECIALTIES

Fluorescence and Phosphorescence

The availability of special lamps emitting long- and short-wave ultraviolet light at a moderate cost has opened up an entirely new field for the mineralogist—fluorescence. The phenomenon was first described in 1852 by Sir George Stokes, who named it after the strongly fluorescent mineral fluorite. Under special lamps made for the purpose, many minerals fluoresce in characteristic colors, ranging from red to blue across the entire visible spectrum. A few minerals that fluoresce strongly in individual colors are the following:

calcite, corundum, tremolite—red
anglesite, barite, chondrodite—orange
scapolite, sphalerite, zircon—yellow
aragonite, autunite, willemite—green
fluorite, celestite, scheelite—blue

Not all specimens of a given species will fluoresce, for as a rule the presence of some impurity is required as an activator. The most important activator of fluorescence in minerals is believed to be manganese.

The sources of flourescent minerals are spread across the earth, and the possibility of a specimen's responding to ultraviolet irradiation cannot be predicted for specimens from any one source; thus, actual tests with an ultraviolet lamp must be made. A few localities produce pieces that fluoresce strongly—for example, wernerite from Quebec, Canada, and the large number of outstanding species from the world-renowned localities at Franklin and Sterling Hill, New Jersey. The property of fluorescence first discovered in minerals is now adapted for use in fluorescent lighting and television screens.

Some minerals will continue to emit light after the ultraviolet rays are removed. Such minerals are said to phosphoresce—that is, they continue to glow like phosphorus, an element that glows in the dark after exposure to the ultraviolet light of the sun.

Sand Collecting

Black sands often contain surprises for the collector, including gold and small gemstones. The most abundant minerals in these sands are magnetite, ilmenite, garnet, amphiboles, and pyroxenes; a number of others may be present to a smaller extent. A polarizing microscope must be used to determine the presence of many of them, unless they occur in good crystals, which is not frequent. The magnetite,

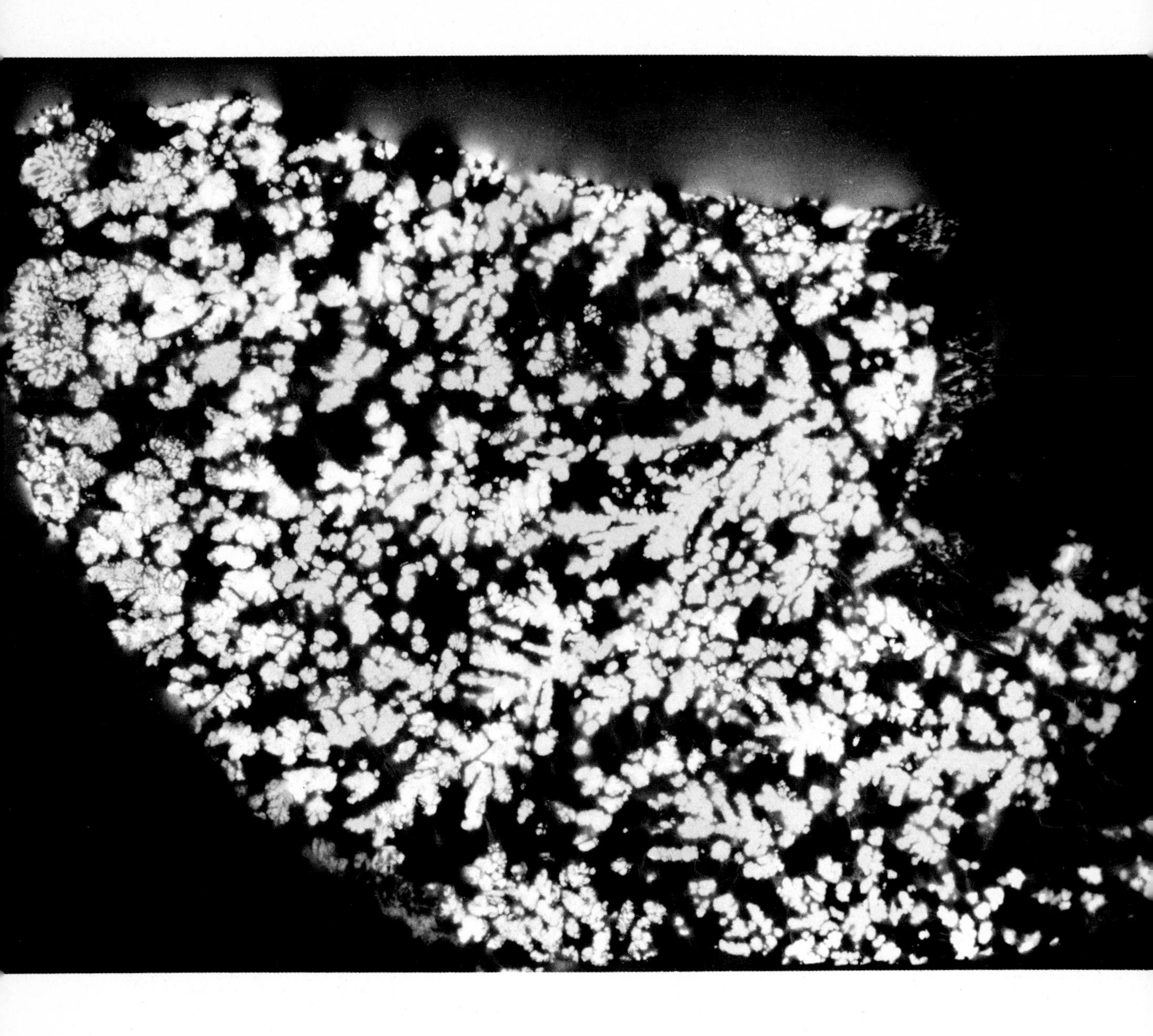

however, often the major component of the sand, can be easily separated with a simple magnet.

Radioactive Minerals

Radioactivity was discovered accidentally in 1896 by Henri Becquerel, as a result of his experiments concerning fluorescence. The phenomenon can be detected by using a Geiger counter. Another method of detection is to place the suspected mineral on a photographic plate in a dark box for thirty-six hours, after which the plate is processed. If the mineral is radioactive, the parts of the plate that have come in contact with it will be exposed by its rays.

ESSENTIAL FIELD EQUIPMENT

In the field one is continually handling hard rocks that contain the sought-for specimens; seldom are the minerals loose and free to be picked up. To be successful in separating the desired minerals from the waste rock, you must be adequately prepared with proper rock-working tools. Fortunately, comparatively little equipment is essential for the beginner.

Autoradiograph of uranium (Ruggles Mine, Grafton Center, N.H.) made by exposing a photographic plate to the radioactive mineral.

The most important item is a suitable hammer. This should be a typical lightweight geologist's pick, with either a replaceable wooden handle securely wedged in the eye of the hammer, or a fixed metal handle with leather grip. Geologist's picks are made of hardened steel and are not desirable for use in striking chisels.

A second item of importance, for working sedimentary rocks, is the chisel-edged bricklayer's pick. Next comes a cold chisel with either a broad or a narrow edge, and a moil, useful when considerable heavy work is necessary to free important specimens. The chisel should be struck with a two- or three-pound striking hammer, which is hard enough to withstand such rough work and tough enough to prevent fragments of the hammer from breaking away and injuring the user and others nearby. The hammer, however, is too soft for direct work on large, hard rocks. For this, the essential tool is a five- or six-pound stonemason's hammer fitted to a thirty-inch handle.

In addition to these hand tools, the professional collector may occasionally find it necessary to use explosives to release and remove rock containing desired minerals. However, the amateur should *never* handle high explosives, for improper use can be extremely dangerous and even fatal.

A knife of good steel will be needed to test the hardness of pieces of mineral.

A powerful magnet will be used for testing material for magnetic properties and for separating magnetic minerals from nonmagnetic species. A magnetized knife blade can be used more skillfully to pick up small magnetic fragments.

A scriber is useful for testing hardness. Though it has about the same hardness as the usual knife blade, its sharp point gives it an advantage, since it enables the collector to test small surfaces with ease and greater certainty. When not in use, the point can be removed, reversed, and held securely in the chuck for protection.

A pocket magnifier is almost as indispensable as a hammer or a knife. The kind that folds into a handle is satisfactory for a quick survey of large areas on a specimen; however, for a careful and more exact examination, the triplet magnifier of 10X is greatly superior.

For special collecting, the serious collector may want to carry, in addition, a small miner's pick (weighing about three and a half pounds) and a light, short-handled shovel for loosening and removing decomposed rock. Excellent crystals and often rare minerals occur in weathered and disintegrated schists, marbles, pegmatites, and granites. A screen for separating the coarse and fine material is useful. A gold pan is often desirable for washing away low specific gravity minerals and concentrating the heavy ones.

Once you have retrieved your specimens from the ground, you'll need a container in which to carry them home. Almost any kind of bag or basket will do, as long as it is strong enough, particularly in the straps and handles, to withstand the weight of the specimens. If you will be doing much collecting, you will find it worthwhile to buy a stout canvas or duck bag, with leather reinforcement on the lower half and a wide shoulder strap. An army knapsack from the surplus store makes an excellent carrier. If no bag is available, plenty of newspaper and some stout cord will do; to protect the specimens, wrap each one separately before tying the individual packages together. A strong canvas hunting coat with its typical large pockets is a good substitute for a collecting bag.

Unless specimens are well wrapped when placed in the bag, they are sure to be damaged by abrasion. A few layers of newspaper around each specimen will preserve them. Very fragile specimens need special protection: tissue paper or cotton batting, or even small individual cartons, tins, or plastic boxes. It pays to protect valuable mineral specimens adequately at all times, for the value of a bruised specimen decreases rapidly. A long and important trip can be reduced to failure by carelessness in handling collected material.

Record Keeping

As it is easy to forget exactly where specimens are collected, it is important to carry along a notebook and pencil, maps, field labels, streak plate, pieces of sharp-edged quartz, and a camera. Numbers should be added to a specimen before leaving the collection site. A small

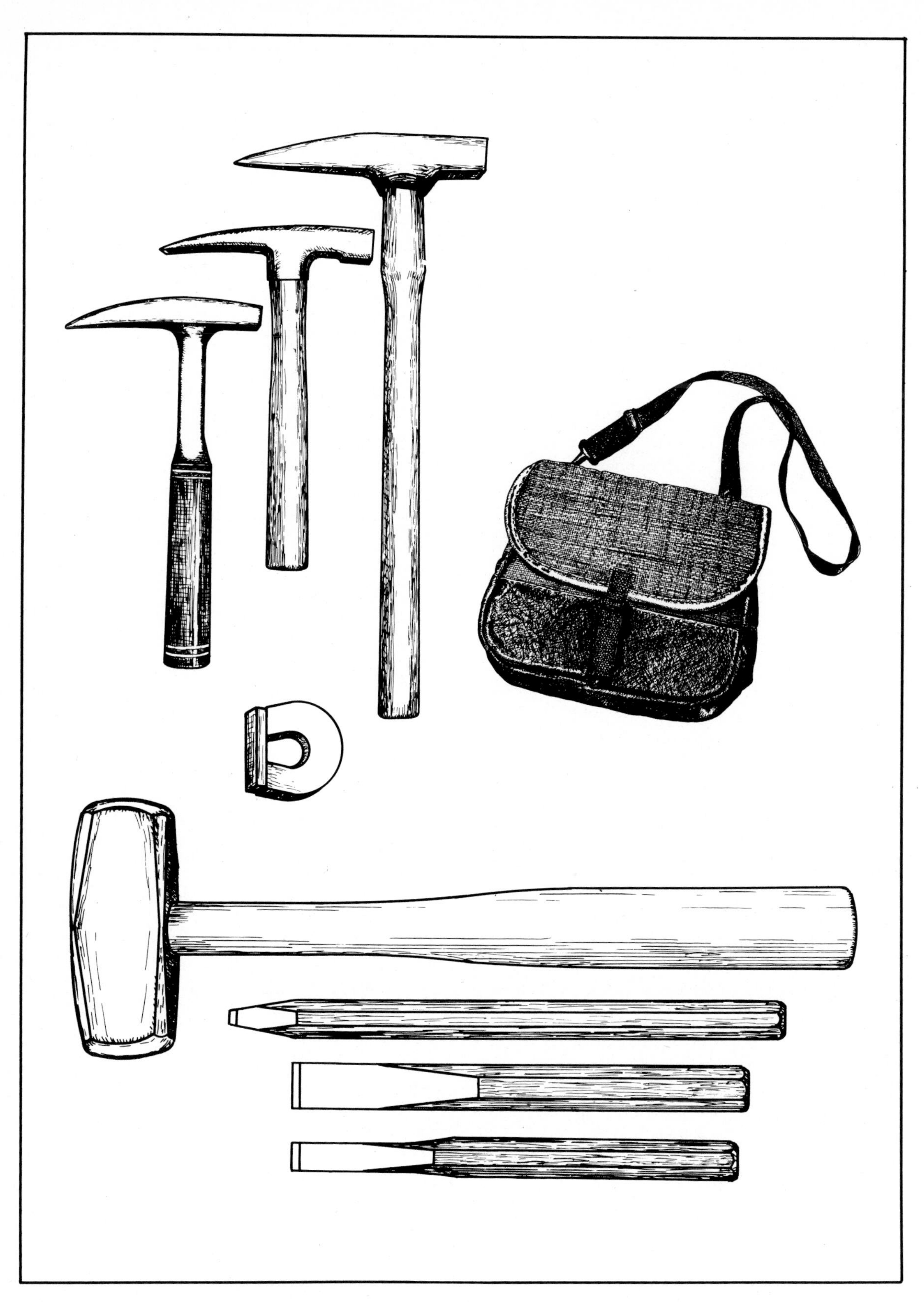

Basic equipment for working in the field. *Top left:* Two geologist's picks and a bricklayer's pick (middle). *Top right:* Leather-reinforced canvas bag, for carrying collected specimens. *Center:* Large magnet. *Above:* A striking hammer, a moil, and two cold chisels (bottom).

roll of tape that can be written on makes a good label that can be applied quickly without damaging the specimen.

Specimens should be recorded by number, with all pertinent data: the date, the field number, the probable name of the mineral, name of the mine or quarry, geographical location, part of mine where collected or if from the dump; if more than one dump occurs, the name of the specific one or its location. If no notebook is at hand, the data can be written on a piece of newspaper and wrapped with the specimen. If a fellow collector gives you a mineral, be sure to get all the data on the specimen. A permanent record will be invaluable in the future for most collectors.

Photographing mineral specimens is an additional method of record keeping. For the average collector, a 35-mm camera is recommended. Photographs made in the field before specimens are dug up are an excellent way of recording their occurrences visually. They are especially valuable when, for various reasons, a mineral cannot be collected and must be left in its original site. (Of course, all information should be written down as well, so that the photographs can be labeled.) Photographs of specimens in the home collection, using photographic lights and various colored backgrounds, can be developed to a fine art; indeed, mineral photography is a highly developed branch of the mineral hobby that is enjoyed by many collectors for its own sake.

CARE AND STORAGE OF YOUR COLLECTION

Many collectors live in small homes, apartments, or trailers, where storage space is at a premium. In such limited quarters, specimens three to five inches in size can quickly fill all available space, leaving room in the collection for only a few species. Here the small microcrystals should be most welcome, for hundreds of them can be stored in the space that only a few single, large specimens would occupy. The larger, showy minerals can then be used here and there as superb accents. Another advantage of microcrystals is that they have the most perfect form of all crystals; when viewed with a microscope, they provide a real delight that far surpasses the pleasure of the average hand specimen, except for purposes of display.

There are as many ways of storing a mineral collection as there are mineral collectors. Minerals are too valuable to be neglected and too difficult to replace for one to be careless about their good keeping. Almost any type of storage facility is adequate, provided that the mineral specimens are separated inside by boxes or partitions and are not allowed to roll against each other.

Many collectors prefer a wooden cabinet with drawers of various sizes; some are made specifically for mineral specimens. The minerals may be stored in the drawers in shallow paper trays or boxes, or partitioned plastic or wooden boxes, or laid out on felt or cotton wadding. Sometimes specimens are wrapped individually in plastic bags. Each bag should contain complete information about the specimen, its identity, where collected, associated minerals, type of deposit from which obtained, when obtained (and from whom), whether purchased, and the price paid. Every specimen should have a catalog number attached to it, according to the record-keeping system used by the collector. A number should be attached to both the specimen and the information label, so that the vital data about the specimen can never become lost.

A quick and effective way to number a specimen is to type the number on a medium-weight paper, trim the number closely, apply a small dab of Duco cement to the most undesirable part of the specimen, apply the number over the dab, and then apply a coat of cement over the number. The specimen can later be repeatedly washed or cleaned in a sonic cleaner without defacing the number. Minerals soluble in water should of course not be washed.

Iron stains can often be removed from nonsoluble specimens by a concentrated solution of oxalic acid. More stubborn stains can be conquered with a 10 percent solution of muriatic acid if the mineral is not soluble in this acid. Ultrasonic devices can be very effective in cleaning minerals, if they carry grinding and polishing powders, and in removing dirt from cavities, cracks, and rough, pitted surfaces.

7.
Minerals Described, Minerals Used

There are many ways of classifying minerals in order to describe them. Any classification scheme is basically for convenience. However, there are certain properties and characteristics of minerals that suggest certain classifications as more logical than others.

The problem is a practical one for museums, universities, and mineral collectors, who must physically arrange their specimens in some kind of order. Some collectors use an alphabetic system. However, most large collections are arranged according to a chemical classification. This is a matter of historical precedent, based on the work of James D. Dana as published in his *System of Mineralogy* (1850). This book set forth a classification of minerals according to their chemistry, starting with simple elements—oxides and sulfides—and proceeding to more complicated compounds. The so-called Dana system has been used widely for more than a century.

However, our appreciation of minerals is based on their utility as well as their beauty and fascination. Minerals are the sources of metals and most of the other resources that form the basis of civilization. Standards of living worldwide have been increased through the efficient use of mineral resources. Over a period of several thousand years man has discovered more and more new metals contained in various minerals, and more and more new uses for minerals *per se*.

Now, more than ever before, we must regard metals as nonexpendable resources. We

have seen that nature required very long times to concentrate metal-bearing minerals into deposits that are considered economically profitable to mine. Once exhausted, such deposits cannot readily be replaced. Refined metal scrap should therefore be set aside for reprocessing, so that future generations may share the resources we now lavishly enjoy.

The following descriptions of minerals are arranged according to the principal metals or elements they contain. Of course, a mineral may be the source of several metals (or be composed of more than one element), but usually one metal is of dominant value. Only a few minerals have been selected for each metal. They are ones well known to collectors, and are frequently available for purchase or can be found in the field. The metals or elements are arranged in alphabetical order; their names appear in capital letters along with their chemical symbols. Following each metal are entries for the minerals that are its significant sources.

To know a mineral, even in a modest way, we must know something about its crystal system and class, its chemical composition, and its physical properties—color, luster, streak, hardness, specific gravity, cleavage, fracture, and tenacity. In many cases, a mineral may be identified solely on the basis of these characteristics, given here in the mineral's description. In other cases, the collector must make some simple chemical and blowpipe tests. Usually a few basic tests will reveal a mineral's identity. The occasional specimen that is more difficult to identify may be tested for its chemical composition, indicated by its chemical formula—stibnite, for example, is composed of antimony and sulfur, as revealed by its formula, Sb_2S_3. A group of tests for the important chemical elements of which minerals are composed is given in the appendix. They should prove sufficient to determine the identity of all but the most difficult specimens. For a list of the chemical elements and their symbols, consult the table at the back of this book.

Finally, significant clues to a mineral's identity may be derived from the types of deposits in which it occurs. The occurrences listed here, with representative geographical locations, will also guide the collector in planning trips in the field.

Abbreviations used

BB:	before the blowpipe
C:	cleavage
CT:	closed tube
F:	fusible
Fr:	fracture
H:	hardness
L:	luster
OF:	oxidizing flame
OT:	open tube
RF:	reducing flame
S:	streak
SG:	specific gravity

Pages 124–125: China jug with cobalt-blue decoration, box of cobalt metal, and piece of smalite, an ore of cobalt, from Cobalt, Ontario. *Left:* From Inyo Province, Japan, striated prisms of stibnite, an ore of antimony, used with lead in alloys, such as pewter.

ALUMINUM (Al)

Aluminum was discovered in 1827, but as late as 1852 the price of the metal was still $545 per pound! In 1886, Charles M. Hall, a 23-year-old student at Oberlin College, became interested in aluminum metallurgy. His crude experiments led to the discovery of an inexpensive process for extraction, and today the price of aluminum is about thirty cents per pound. Metallic aluminum does not occur in nature. The metal is extremely light (SG 2.7), ductile, malleable, and a superb conductor of electricity. Numerous alloys with many other metals are produced today. Some have the tensile strength of steel with only a third the weight. Aluminum is very easy to fabricate and has hundreds of uses, including transportation, electrical transmission, recreational vehicles, appliances, office equipment, and packaging.

More than 150 minerals contain aluminum as an important constituent. Bauxite, a complex mixture of hydrated oxides, is the chief ore today. Cryolite—sodium aluminum fluoride—is important in aluminum refining because molten cryolite will dissolve bauxite; this was the key in Hall's refining process. Diaspore and gibbsite are important clay minerals; both are simple aluminum oxides.

Bauxite

Named for Les Baux, France, where it was originally found and described, bauxite, $Al_2O_3 \cdot 2H_2O$, is essentially a rock containing a massive, earthy, compact material resembling a hardened clay of a somewhat porous texture; it has a dull, earthy luster. In the absence of impurities the color is gray or white but it is usually tinted brown, brick red, or ocher yellow by a small amount of iron. Bauxite is fine-grained and opaque. Other physical properties are extremely variable.
Tests: CT yields water. Distinguished by tests for aluminum.
Occurrence: In Arkansas, bauxite occurs as an alteration product of a nepheline syenite. The most important bauxite deposits of the world rest upon clays resulting from clay-bearing limestones or rocks altering to clay. Field work indicates that the bauxites originated from the clays by the removal of silica, through prolonged tropical or subtropical weathering.

Commercial deposits are present in many tropical countries, notably Surinam, Jamaica, French Guiana, and French Guinea; in the United States in Georgia, Arkansas, Alabama, and Florida; along a belt from southern France eastward to Yugoslavia; and in Australia and the East Indies. Its most important use is as an ore of aluminum.

ANTIMONY (Sb) *(illus. p. 137)*

This interesting element is widely used in low-melting-point alloys, as a constituent of pewter, and in the metals used for making type. Antimony expands when it cools, so its alloys will fill all the cracks and irregularities in a mold, thus producing clear, sharp type.

About fifty-five minerals contain antimony. Many of these are admixed in nature and many are antimony ores. Native antimony occurs in nature, but is very rare. Some of the common oxidation products are called "antimony ochers," and include valentinite, cervantite, stibiconite, senarmontite, and kermesite. The antimony mineral most familiar to collectors, however, is stibnite.

Stibnite *(illus. pp. 127, 137, 161)*

Orthorhombic, bipyramidal; antimony sulfide, Sb_2S_3. In long prismatic crystals, deeply striated vertically and steeply terminated with many faces. Crystals often bent; even large crystals can be bent if care is exercised; also occurs in small, radiating groups of acicular crystals often having a bladed structure; sometimes as granular masses. Name derived from the Latin for antimony, *stibium,* "that which marks." Dark grayish black; L metallic, brilliant; S metallic; H 2; SG 4.63; C {010}; Fr subconchoidal; brittle; slightly sectile; flexible but not elastic.
Tests: Easily fused in a candle or match flame; BB colors the RF greenish blue; yields dense white fumes with the odor of sulfur dioxide and a white sublimate on charcoal near the assay; soluble in HCl; fragments give a characteristic yellow coating in KOH; OT gives white, nonvolatile sublimate below a white volatile sublimate; CT a weak ring of sulfur over a red sublimate of antimony oxysulfide.

The crystal habit, perfect pinacoidal cleavage, dense, white fumes with sulfurous odor, low fusibility, color and softness are characteristic.
Occurrence: Usually in low-temperature deposits, associated with galena, sphalerite, gold, silver, cinnabar, pyrite, and other antimony minerals. The finest specimens come from Iyo Province, Japan, and Hollister, California; radiating groups from Romania and at Manhattan, Nye County, Nevada; Coeur d'Alene district, Idaho; Hunan, China; Australia; Bolivia; Peru; Mexico; Algeria; and the Balkan countries.

ARSENIC (As)

Arsenic is a white metal that rarely occurs uncombined in nature. The metal tarnishes readily; H 3.5; SG 5.63 to 5.78; C perfect basal. This element has been known since ancient times. Its minerals are usually found in low-temperature deposits, associated with minerals of silver, copper, nickel, cobalt, and lead. Arsenic is an essential element in more than 165 minerals, many of which are found in splendid mineral specimens.

Arsenopyrite

Monoclinic, prismatic; a sulfide

and arsenide of iron, FeAsS. Cobalt, nickel, and bismuth may be present in small amounts. Metallic white inclined to white, metallic gray; L metallic; S dark grayish black; H 5.5 to 6; SG 6 to 6.2; C {110} distinct; Fr uneven; brittle.
Tests: BB on charcoal fuses at 2; RF becomes magnetic; gives a sublimate of arsenious oxide, As_2S_3; fumes have sulfurous and garlic odors; OT gives arsenious oxide and acid reaction; OT arsenic mirror and red sublimate of arsenic sulfide. Crystal forms characteristic.
Occurrence: Found in intermediate and deep-seated deposits with silver minerals, galena, sphalerite, pyrite, chalcopyrite, tetrahedrite, and minerals of tin, cobalt, nickel, and tungsten. Some localities are in Saxony, Yugoslavia, Sweden, England, New Jersey, Ontario, also in Joplin, Missouri, and Oxford County, Maine, and many others. An ore of arsenic.

Loellingite

Orthorhombic, bipyramidal; iron diarsenide, $FeAs_2$. Occasionally may contain some cobalt and nickel. Metallic white to light metallic gray; L metallic; S grayish black; H 5 to 6.5; SG 7.4 to 7.5; C {100}, {010} distinct; Fr uneven; brittle.
Tests: BB on charcoal gives a magnetic globule; F at 2; dense white fumes with garlic odor; coats charcoal with a white sublimate; CT gives arsenic mirror; OT gives oxide sublimate. Resembles arsenopyrite in crystal form and appearance but lacks sulfur. Indistinguishable from skutterudite and safflorite without chemical or X-ray tests.
Occurrence: Intermediate- to high-temperature deposits associated with iron and copper sulfides, also found in silver and gold deposits, sometimes in pegmatites. Found at Lölling, Austria; Harz Mountains, Germany; good crystals from Norway and Sweden; as an inter-

Below: Beryllium (right), found in aquamarine (left) and beryl (center); used in alloys to increase elasticity, as in the beryllium-copper spring.
Bottom: Azurite (Morenci, Ariz.)—decorative stone, copper ore, and a source of blue pigment.

growth in garnet in a pegmatite at Center Strafford, New Hampshire; also in Ingersoll Mine, Keystone, South Dakota. An ore mineral of arsenic.

Orpiment

Monoclinic, prismatic; arsenic trisulfide, As_2S_3. Usually in aggregates with reniform or botryoidal shapes, also in columnar, foliated, fibrous, and granular masses. The name is derived from the Latin *auripigmentum,* "golden paint," in allusion to the brilliant yellow color. Bright lemon yellow; L resinous to subadamantine; S yellowish; H 1.5 to 2; SG 3.49; C {101} eminent, {100} in traces; cleavage laminae flexible but inelastic; transparent to translucent; sectile.
Tests: BB on charcoal burns with a blue flame, emitting a sulfurous odor and that of garlic; F 1; white sublimate on charcoal; OT gives a white sublimate of As_2O_3 and acid reaction with moistened litmus paper; CT gives a dark yellow sublimate. Its brilliant color, eminent cleavage, inelastic, flexible cleavage laminae, and pearly luster distinguish it from sulfur and all other yellow minerals.
Occurrence: A low-temperature mineral of the upper hydrothermal veins and a sublimation product of fumaroles. Excellent crystal specimens come from Mercur, Utah; in loose masses of hairlike fibers of great beauty from Manhattan, Nevada; Hungary; and Macedonia. An ore mineral of arsenic.

Realgar *(illus. p. 81)*

Monoclinic, prismatic, sometimes in excellent crystals of short, prismatic habit; also as fine to coarse granular compact masses and as incrustations. Name derived from the Arabic *rahj al-ghār,* "powder of the mine." Arsenic monosulfide, AsS. A deep aurora red, changing on exposure to light to orange; L subadamantine to resinous; S light red to orange red; H 1.5 to 2; SG 3.56; C {001} and {010} fair; Fr small conchoidal; brittle; transparent to translucent.
Tests: BB on charcoal burns with a blue flame, emitting sulfurous and garliclike odors and a white sublimate. OT gives a white sublimate and acid reaction with litmus paper; CT gives a red transparent sublimate; decomposed by HNO_3 and soluble in caustic alkalies. Characterized by aurora red color, softness, and lower SG than most red minerals.
Occurrence: Intermediate- and low-temperature veins; associated with minerals of antimony, silver, gold, and mercury. Fine crystals from Hungary, Macedonia, Switzerland, and Japan. In the United States beautiful specimens from Homestead Gold Mine, Lead, South Dakota; and Manhattan, Nevada, and Mercur, Utah. An ore mineral of arsenic.

BARIUM (Ba)

The name itself means "heavy." Metallic barium is seldom used as a metal; its primary uses involve the minerals in which it is found. It is an essential element in only about twenty-five minerals.

Barite *(illus. pp. 36, 86, 117, 130)*

Orthorhombic, bipyramidal; crystals are usually tabular parallel to the base, or prismatic and usually very simple, although at times the groups are highly complex, as in the aggregates known as "desert roses." Also occurs as nodules, stalactites, laminated and granular masses. Barium sulfate, $BaSO_4$, may contain small amounts of strontium and calcium. White, yellowish, brown, gray, greenish, blue, or red, sometimes colorless transparent; L vitreous; S white; H 2.5 to 3.5; SG 4.3 to 4.6, which is high for a nonmetallic mineral; C {001} perfect, {110} good; Fr uneven; brittle.
Tests: BB on charcoal decrepitates; F 4, colors flame yellowish green; fused with calcium carbonate on charcoal; when as-

Top: Barite, used in white paint and in oil drilling. *Center:* Fire-sprinkler head and firedoor lock, made from a low-melting-point alloy of lead (on galena, top r.), cadmium (lower l.), and bismuth (lower r.). *Above:* Borax, from California, a mineral with almost infinite uses.

say is placed on a silver coin and moistened, leaves a black stain of silver sulfide; insoluble in acids. High SG, lack of effervescence in acid, and three cleavages distinguish it from calcite, dolomite, magnesite, and similarly appearing minerals; distinguished from feldspar by inferior hardness.

Occurrence: Barite is one of the common minerals in the intermediate- and low-temperature fissure veins and in other deposits formed under similar conditions, often abundantly associated with minerals of lead, copper, zinc, and silver; it occurs as nodules in residual clay derived from weathering limestones, and as fine-grained lenses in sediments probably of contemporaneous origin and sometimes of large size.

Arkansas and Missouri are the leading producers in the United States. Transparent crystals weighing up to one hundred pounds have been found at Frizington in Cumberland, England. Fine crystals come from Freiberg, Germany; Felsöbanya, Romania; and Cheshire, Connecticut; delicate blue crystals from Sterling, Colorado; brown and sherry yellow crystals up to 4.5 inches long are found in cracks with yellow calcite in septaria in the area north of Wasta, South Dakota. Other localities are Norman, Oklahoma; Palos Verdes, California; and Rosiclare, Illinois.

Uses: Barite is used in the manufacture of lithopone, a pigment that produces a brilliant, white, opaque paint. The largest market for barite is in the oil and gas industry, where it is ground and mixed with water; the heavy fluid thus formed is used in rotary drilling to control gas pressure and to prevent drill-hole cave-ins. Barite is also an ore mineral of barium metal.

Witherite

Orthorhombic, bipyramidal; always cyclically twinned on the prism {100} producing pseudohexagonal pyramids; also globular, botryoidal, columnar, granular in coarse fibers. Named for William Witherington, an English physician and mineralogist, who first called attention to the mineral. Barium carbonate, $BaCO_3$; a small amount of strontium and calcium are sometimes present. Colorless to white, gray, brown, green, or yellow; L vitreous to resinous; S white; H 3 to 3.5; SG 4.29; C {010} distinct, {110} imperfect; Fr uneven; brittle; transparent to translucent.

Tests: F 2; BB colors the flame yellowish green; soluble in dilute HCl; addition of H_2SO_4 produces a white precipitate of barite. Characterized by cyclic twinning, producing pseudohexagonal crystals and a yellowish green flame.

Occurrence: Low- to medium-temperature veins of galena, sphalerite, calcite, and fluorite. The original locality is at Alston Moor, Cumberland, England; also from Bohemia and France. Of rare occurrence in the United States: in massive form from near Platina, California; near Lexington, Kentucky; as rounded, twinned crystals in the fluorite veins in Minerva Mine, Rosiclare, Illinois; and near Thunder Bay, Ontario. An ore mineral of barium and a source of barium salts.

BERYLLIUM (Be) *(illus. p. 129)*

Beryllium oxide was discovered in 1797, and the metal was isolated in 1899. However, only in the twentieth century did technology find uses for this valuable metal. Beryllium is hexagonal, grayish white in color; SG 1.80, H about 6. The melting point is high, about 1350°C, which is twice that of aluminum. Beryllium is more rigid than steel, a good electrical conductor, and nonmagnetic. Its principal uses are in alloys with copper, iron, nickel, and chromium. Even when present in amounts of only a few percent, beryllium increases the strength, elasticity, and corrosion resistance of copper. Springs made

Top: Ulexite (California), an ore of borax. *Center:* Crocoite (Tasmania), a rare ore of lead and chromium and a desirable collector's item. *Above:* Chromite (Webster, North Carolina), a source of chromium, which is used in stainless steel, as in this stainless steel bar and pitcher.

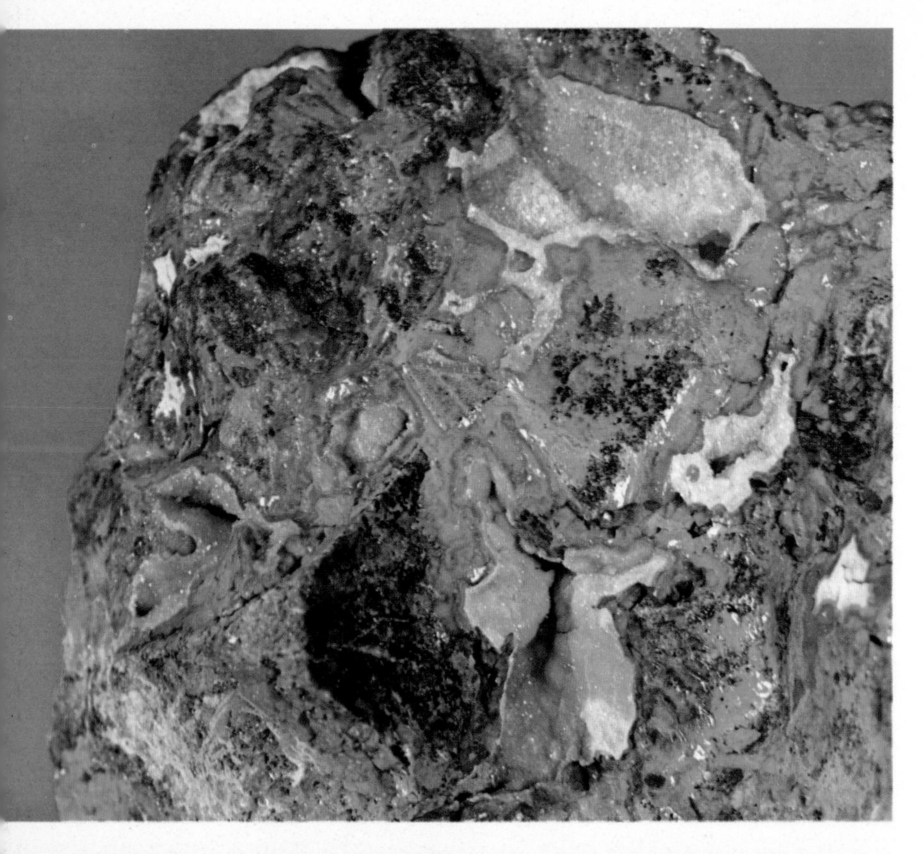

Top: Chrysocolla, a common mineral in copper deposits, sometimes used as a gemstone (Globe, Arizona). *Above:* Malachite, from Lavender Pit in Bisbee, Arizona, commonly associated with azurite and other copper minerals, also valued as a decorative stone and used in jewelry.

of copper-beryllium alloy are used widely today. The widespread use of beryllium is restricted only because of its scarcity and high cost.

Beryllium occurs as an essential element in only about twenty minerals. By far the most common and most important is the mineral beryl, although other species, such as beryllonite and herderite, are popular among collectors.

Beryl *(illus. pp. 10, 83, 129, 179, 181)*

Hexagonal, bipyramidal; a beryllium, aluminum silicate, $Be_3Al_2(SiO_3)_6$. Sodium, lithium, or cesium may substitute for beryllium, thus lowering the commercial value of the mineral. White, greenish white, clear colorless, rose red, yellow to orange, pale to bright grass green, bluish green to blue; L vitreous; S white; H 7.5 to 8; SG 2.69; Fr conchoidal to uneven; cleavage indistinct basal; brittle; transparent, translucent. *Tests:* BB becomes clouded after prolonged heating; at very high temperature forms a slag; F 5.5; borax bead colorless except for the green variety, which yields a green glass; not acted upon by acids. Characterized by hexagonal crystals, conchoidal fracture, and color; distinguished from apatite (H 5) and from topaz, which has a perfect basal cleavage.
Occurrence: Beryl is a typical pegmatite mineral usually associated with the quartz-rich phases of pegmatites, sometimes in huge crystals as much as six feet between opposite faces, over thirty feet long, and weighing up to seventy-eight tons. (A crystal of this size was removed from the Bumpus Quarry at Albany, Maine.) The localities producing good crystals are numerous; among the most important are the Ural Mountains, Madagascar, eastern Brazil, South-West Africa, India, Maine, Connecticut, North Carolina, the Black Hills of South Dakota, and California. Emerald—the green variety—comes

from Colombia and the Ural Mountains. Commercial deposits of beryl occur in Argentina, Brazil, Mozambique, the Union of South Africa, and India.
Uses: The use of beryl for gems is described in Chapter 8. The clear, colorless varieties of beryl, when associated with colored tourmalines, spodumene, lepidolite, cassiterite, and columbite, are often accompanied by the rare mineral pollucite, which is the most important ore of cesium. Beryl is the most important ore mineral of beryllium.

Herderite

Monoclinic, prismatic; in stout crystals, also in spheroidal or radiating aggregates, fibrous and in grains; twin plane {001}; named for S. A. W. von Herder, a mining official in Freiberg. A basic phosphate of calcium and beryllium, $CaBePO_4(F,OH)$, containing up to 5.58 percent beryllium. Pale yellow to greenish white; L vitreous; S white; H 5 to 5.5; SG 2.95 to 3; C {110}; fracture subconchoidal; brittle.
Tests: BB fuses with difficulty and becomes white and opaque; CT intensely ignited yields acid water; soluble in acids.
Occurrence: A pegmatite mineral of the late stages of low-temperature crystallization; found in Saxony and Bavaria, Germany; Mursink, Ural Mountains, USSR; in Stoneham, Newry, and Paris, Maine; and Fletcher and Palermo Mines, North Groton, New Hampshire.
Use: Too rare for an ore mineral of beryllium; a collector's item.

BISMUTH (Bi) *(illus. p. 130)*

Bismuth has been recognized as a metal for more than two hundred years, but was long confused with tin. Native bismuth rarely occurs in nature. Its primary uses are in alloys with cadmium, lead, mercury, antimony, and indium, for low-melting-point alloys used in safety devices, such as sprinkler systems and temperature regulators. The metal melts at only

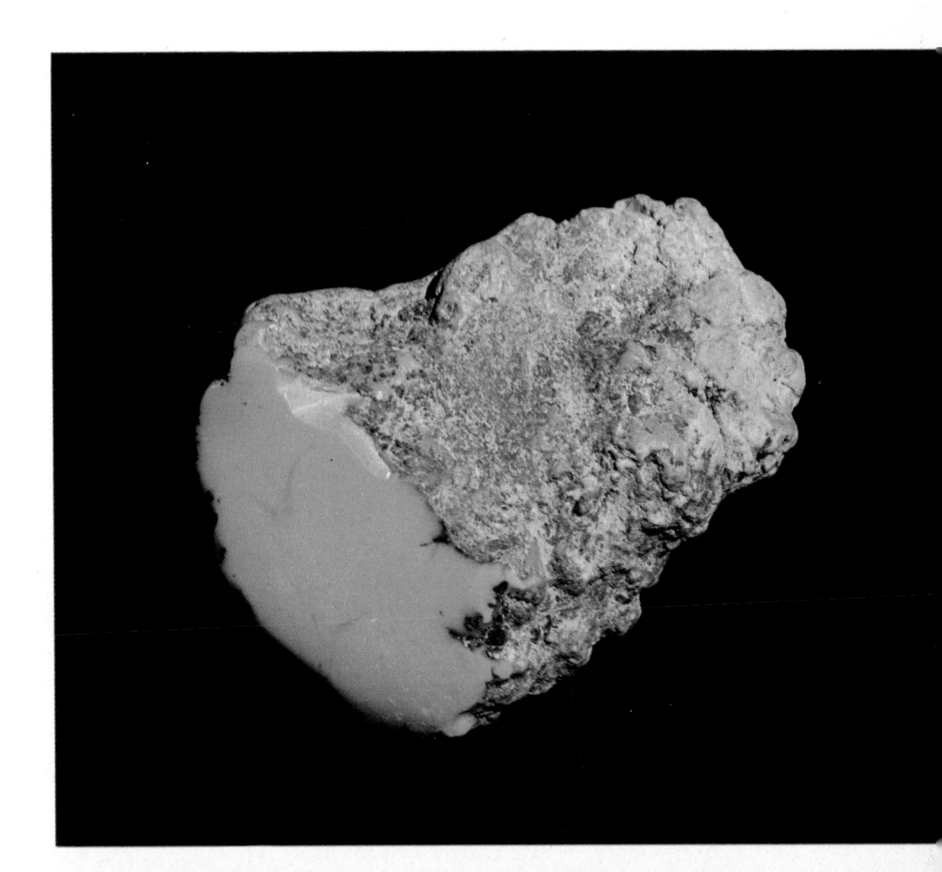

Top: Turquoise, from Austin, Nevada, a mineral that rarely occurs in crystals but is a popular gemstone in many parts of the world. *Above:* Cast-iron gear and steel bar, with alloying elements (bottom) and their ores (top)—siderite (Connecticut) and oölitic hematite (New York).

From Broken Hill, New South Wales, Australia, fine reticulated crystals on an excellent specimen of cerussite, a common ore mineral of lead.

271° C; H 2 to 2.5; SG 9.75. Bismuth is an essential constituent in only about twenty-five minerals. It is generally recovered as a by-product in the smelting of minerals for other metals.

Bismuthinite

Orthorhombic, bipyramidal, in stout to acicular crystals striated vertically on prism faces; also massive, foliated, or fibrous. Named in allusion to its composition, bismuth trisulfide, Bi_2S_3; may contain small amounts of tellurium, antimony, iron, copper, and lead. Light metallic gray to metallic white with a yellowish or iridescent tarnish; L metallic; S light gray; H 2; SG 6.4 to 6.7; C {010} perfect; two others imperfect; Fr uneven; flexible and slightly sectile; opaque.

Tests: BB F 1; on charcoal yields spherical globule and an orange yellow sublimate, which is yellow when cool; easily soluble in HNO_3; yields a white precipitate on dilution with water; on plaster yields bright scarlet coating with brownish borders. Characterized by its resemblance to antimony but is much heavier and gives tests for bismuth. Alters to other bismuth compounds.

Occurrence: Usually in intermediate-temperature veins associated with native bismuth, arsenopyrite, and sulfides, also in tin and tungsten veins; occasionally found in granite pegmatites associated with bismuth, galena, apatite, and albite. Obtained at Schneeberg, Saxony, Germany; Romania; Italy; France; Sweden; and England. In the United States at Keystone and Custer, South Dakota, and at Haddam, Connecticut. Important deposits are located in Bolivia in the tin-tungsten veins. An ore of bismuth.

BORON (B)

Boric acid was known before the element boron, a nonmetal, was isolated in 1808. Boron melts at 2300° C, SG 3.33 (crystals), and its electrical resistance decreases drastically with increased temperature. Boron is used extensively in making steel because its presence, even in small amounts, confers tremendous hardness. Boron in various alloys increases strength, hardness, and abrasion resistance. Boron nitride and boron carbide, synthetic substances, have a hardness that approaches that of diamond and are used as abrasives. Boron is used in nuclear reactors, and its compounds are useful in detergent mixtures. Boron hydrides make excellent rocket fuels.

About thirty-five borates (compounds with oxygen) and nearly half as many borosilicates have been identified. The most important of these are boracite, borax, kernite, and ulexite. The important borosilicates are datolite, axinite, tourmaline, and dumortierite.

Beta Boracite

The atomic structure of beta boracite is orthorhombic while its crystal forms are those of the isometric paramorph alpha boracite, which is stable above 265°. The usual forms of alpha boracite are retained by the lower temperature form beta boracite. They are the cube, tetrahedron, and rhombic dodecahedron often highly modified. Magnesium chloroborate, $Mg_6B_{14}O_{26}Cl_2$. Some ferrous iron may proxy for magnesium. Colorless, white, gray, yellow, bluish to dark green; L vitreous to subadamantine; S white; H 7 to 7.5; SG 2.9 to 3; C none; brittle; transparent to translucent. Strong piezoelectric and pyroelectric properties.

Tests: BB intumescent; F 2 to a clear bead, yellowish when hot, white when cold; colors the flame green; heated with cobalt nitrate solution yields a deep pink color (magnesium). Dissolves slowly in HCl; heated with copper oxide on charcoal gives an azure blue flame due to copper chloride. Characterized by the crystal forms and hardness when in crystals. Blowpipe and physical tests should identify the mineral.

Occurrence: In beds resulting from the evaporation of marine and lake waters in dry areas; associated with salts of sodium, magnesium, and potassium. As crystals at Stassfurt, Germany; in salt deposits in France and England; at Choctaw salt dome, Iberville, Louisiana; and in Otis, California.

Use: Excellent cabinet specimens of small crystals; not sufficiently abundant for an ore mineral of boron.

Borax *(illus. p. 130)*

Monoclinic, prismatic; crystals usually short, prismatic but some are up to seven inches long; also in tabular forms parallel to the base; in granular crusts and layers in muds of marshes and lakes in arid areas. Name derived from the Arabic *buraq*, "white." A hydrated sodium borate, $Na_2B_4O_7 \cdot 10H_2O$. Contains more than five barrels of water per cubic yard. Colorless to white, grayish, bluish, greenish; L vitreous to resinous, earthy; S white; H 2 to 2.5; SG 1.70 to 1.72; C {100} perfect, {110} less perfect; Fr conchoidal; brittle; transparent to translucent. Soluble in water yielding a sweetish alkaline taste. Borax melts or is completely soluble in its own water of crystallization at 77° C. Crystals effloresce, turn white, and crumble to a powder.

Tests: BB intumesces and fuses easily to a colorless glass. It yields an intense yellow flame but when mixed with powdered fluorite and potassium bisulfate the flame is light green; gives an alkaline reaction, a reddish brown color on moistened tumeric paper, and when dried the color changes to black when moistened with ammonia. Characterized by its solubility in water, unusual taste, crystal forms, and blowpipe reactions.

Occurrence: With other salts formed by evaporation of shallow lakes and marsh waters in arid areas. Deposits of borax are known in Tibet, Kashmir, USSR, and Iran. The southern part of California is famous for its many deposits of boron minerals, especially the boron open pit in Kramer and Searles Lake in San Bernardino County.

Uses: The uses are legion. Borax enters in some manner into every person's daily activities, through the use of cleaning agents, pharmaceutical preparations, and antiseptics to prevent mold and mildew. It is also used in manufacturing coated papers; fireproofing wood and fabrics; cleaning skins and dyeing and treating leather; as a food preservative, a flux for brazing, and a drier for oils and paints; in porcelain enamels for stoves, refrigerators, cooking utensils, and tabletops; in heat resistant glasses; and an essential ingredient in fertilizers for promoting the growth of plants. Boron nitride, with a hardness approaching that of diamond but with better heat resistance, is gaining in use as an abrasive. Boron trichloride is used as an extinguishing agent for magnesium fires.

Colemanite

Monoclinic, prismatic; usually as short prisms and as pseudorhombohedral crystals; also massive, fibrous, columnar, and spheroidal aggregates. Named for William T. Coleman, in whose mine the mineral was first found. A hydrated calcium borate, $Ca_2B_6O_{11} \cdot 5H_2O$. Colorless to white, also grayish white to gray; L vitreous to subadamantine; S white; H 4.5; SG 2.41 to 2.43; C {010} perfect, {001} distinct; Fr even to subconchoidal; transparent to translucent.

Tests: BB decrepitates, exfoliates, and is reduced to a powder; fuses imperfectly; colors flame green; gives reaction for boron with tumeric paper; soluble in HCl when heated, with separation of H_3BO_3 on cooling. Distinguished by the brilliance and complexity of its excellent crystals; the cleavage, fusibility, and exfoliation BB will be further checks.

Occurrence: Important occurrences in borax fields in Kern, Inyo, and Los Angeles counties in California, and in Nevada, often in geodes lined with crystals associated with celestite, ulexite, and priceite. Also in Argentina. Found in dried-up marshes and alkali lakes with halite, thenardite, gypsum, and other borates. An important ore mineral of boron.

Below: Pyromorphite, from Bohemia (Czechoslovakia), associated with lead deposits and valued by collectors for its striking form and color.

Datolite

Monoclinic, prismatic, in good to excellent crystals. Sometimes remarkably clear and transparent; often highly modified with sharply defined faces; crystals vary in size up to several inches across and more or less equidimensional; also radiating, massive granular to compact cryptocrystalline masses with porcelainlike fracture surfaces. Name derived from the Greek *dateisthai,* "to divide," in allusion to the friable nature of coarse granular specimens. A basic orthosilicate of calcium and boron, $Ca_2B_2(SiO_4)_2OH$. Clear and colorless to white, rarely reddish, amethyst or dirty olive green, light yellow green; L vitreous for crystals, porcelaneous for compact cryptocrystalline variety; S white; H 5 to 5.5; SG 2.8 to 3; C none; Fr conchoidal to uneven; brittle; transparent to translucent.

Tests: BB intumesces to a clear glass, coloring the flame a bright green; F 2; CT yields water; gelatinizes with HCl. Distinguished by the complex, greenish glassy crystals and blowpipe tests; colorless grains may be confused with quartz but quartz is much harder.

Occurrence: Datolite is a typical mineral of the intermediate basic rocks, the fine-grained gabbros and diabases, and occurs chiefly in veins and cavities within these rocks. It is associated with calcite, prehnite, epidote, quartz, sphalerite, galena, pyrite, chalcopyrite, diabanite, stilpnomelane, thaumasite, apophyllite, and other zeolites. It is one of the few minerals to capture the boron released during the solidification of the rock magma. Very fine crystals have been found at Westfield, Massachusetts; at Tariffville, East Hampton, New York; and Meriden, Connecticut. Splendid crystals are also obtained from the trap rocks of Bergen Hill, Great Notch, Paterson, and elsewhere in New Jersey. A compact cryptocrystalline variety is associated with the trap rock of the copper district in Keweenaw County, Michigan. Datolite is also found in Germany, Italy, and Norway. It has no commercial use except as an excellent mineral specimen for museums and private collections. Gem collectors have cut gems from the clear, transparent material.

Kernite

Monoclinic, prismatic; in large crystals of nearly equal proportions and as cleavable grains and granular masses. Crystals up to several feet in diameter occur in the Kramer borate district, California. Named for Kern County, Cal., where it was discovered. A hydrous sodium borate, $Na_2B_4O_7 \cdot 4H_2O$. Clear, colorless to white; L vitreous to pearly; S white; H 2.5; SG 1.91 to 1.92; C {001}, {100} perfect; Fr conchoidal to uneven and splintery; brittle; transparent to translucent.

Tests: BB imparts a green color to the flame, swells and fuses

easily to an opaque white mass, then to a clear glass. Readily soluble in acids and in hot water. CT yields much water; gives reaction for boron on tumeric paper. Kernite is distinguished from borax by its greater hardness and cleavage.
Occurrence: In clay sediments of old lakes as large crystals and as fibrous masses, in veins along with other boron minerals in the Kramer district, Kern County, California, where crystals as large as three by eight feet have been found.
Use: Kernite is an important ore mineral of borax, into which it can be converted easily by the addition of water, thereby increasing its weight by 39 percent.

Ulexite

Triclinic, almost invariably in balls up to about four inches in diameter consisting of fine, interwoven fibers, capillary or acicular grains. The structures are often referred to as "cotton balls." Also in spheroidal crusts, rarely in crystals. Named for the German chemist George Ludwig Ulex, who first gave a correct analysis for the mineral. A hydrated borate of sodium and calcium, $NaCaB_5O_8 \cdot 8H_2O$. White, individual grains or crystals colorless; L vitreous for individual grains, silky for the masses; S white; H 2.5; SG 1.95 to 1.96; C {010} perfect, $\{1\bar{1}0\}$ good; Fr uneven; brittle; tasteless.
Tests: BB fuses with swelling to a clear, blebby glass; CT yields water; soluble in acids; slightly soluble in hot water. No other mineral resembles the "cotton balls" with their shining silky fibers.
Occurrence: This mineral occurs in arid regions associated with other borates and the usual minerals of salt playas, lakes, and marshes. It occurs in quantities in the salt marshes and alkali flats of Esmeralda County, Nevada, and the borax region of southern California; also at Windsor and elsewhere in Nova Scotia with gypsum, halite, glauberite, and pickeringite. Ulexite is found in the dry plains region of Chile and Argentina and in the borate deposits in western Kazakhstan, USSR. An ore mineral of borax.

CADMIUM (Cd) *(illus. p. 130)*

Cadmium melts at a low temperature—320° C—and is therefore useful in low-melting-point alloys. It is a silvery white metal, and all the compounds are poisonous. Cadmium is used extensively in rechargeable nickel-cadmium batteries. It is present in only a few minerals.

Greenockite

Hexagonal, dihexagonal pyramidal; in hemimorphic pyramidal crystals the upper forms are more prevalent and complex, often horizontally striated. The mineral usually is late to crystallize and occurs as a thin crust or film on the associated minerals. It is rare that the crystals are large enough to be distinguishable without the aid of magnifiers. Named for Lord

Stibnite (top l.) from Mexico and galena (top r.), from Joplin, Mo., the chief ores of antimony and lead, respectively; with antimony metal (bottom l.), moveable type made of lead and antimony, and two lead weights.

Greenock. A sulfide of cadmium, CdS. Various shades of yellow and orange; L adamantine to resinous, also earthy; S orange yellow to brick red; H 3 to 3.5; SG 4.8 to 5; C {11$\bar{2}$2} distinct, {0001} imperfect; Fr conchoidal; brittle.
Tests: BB infusible; volatilizes easily in RF; yields a brown sublimate and an iridescent tarnish; CT becomes carmine when hot but changes to yellow upon cooling; soluble in concentrated HCl with evolution of hydrogen sulfide. Its occurrence and blowpipe reactions distinguish it from other minerals.
Occurrence: Forms a yellowish coating on zinc minerals, especially sphalerite. Rarely in cavities in igneous rocks. In Scotland with prehnite, natrolite, and calcite; also found in Bohemia and Greece. With prehnite and zeolites at Paterson, New Jersey, and Friedensville, Pennsylvania; also at Franklin, New Jersey, and around Joplin, Missouri; coatings on sphalerite and smithsonite at many European localities; in the tin-silver mines of Bolivia.
Use: The principal ore mineral of cadmium, never concentrated sufficiently to be exploited alone; it is always obtained as a byproduct in smelting other ores.

CARBON (C)

The occurrence of carbon in nature is widespread, chiefly in organic materials and carbonates of calcium, magnesium, and iron in sedimentary rocks. There is abundant carbon dioxide in the atmosphere. Carbon chemistry, in fact, is the basis of all life on earth.

The carbonate minerals all contain the so-called carbonate group, CO_3, in which a carbon atom is surrounded by three oxygen atoms. Carbonate rocks account for the vast bulk of nonorganic carbon or gaseous carbon dioxide in the earth. Carbon does occur in nature as two very dissimilar minerals: graphite and diamond.

Graphite

Hexagonal, dihexagonal-bipyramidal. Crystals are small, hexagonal plates parallel to the base, {0001}; trigonal striations on base; often poorly radiating and sometimes roughly columnar; scaly, granular to compact. Twinned on a second order pyramid. Name derived from the Greek *graphein,* "to write," alluding to its use for writing. It is composed of carbon and may contain impurities such as iron oxides and clay. Black to grayish black; L metallic, sometimes dull to earthy; S black to grayish black; H 1 to 2; SG 2.09 to 2.23; C {0001} perfect, very easy; flexible but nonelastic; greasy feel; sectile; deep blue by transmitted light in extremely thin flakes, otherwise opaque. A conductor of electricity. Resistant to the action of many chemicals; high vaporization temperature of 3500° C.
Tests: BB infusible, when fused with KNO_3 in a platinum spoon it deflagrates and forms potassium carbonate which effervesces with acids. Graphite is unaltered in acids. It closely resembles molybdenite. Both mark paper but the streak of molybdenite is more metallic in appearance. A pencil mark drawn across the streak of either mineral will readily identify it as being the same or different; in the latter case the mineral is molybdenite. BB molybdenite is reactive while graphite is inert.
Occurrence: Graphite is a typical metamorphic mineral developed chiefly from carbonaceous remains in sediments. Some coal beds—like those in Rhode Island and Sonora, Mexico—may have been changed into graphite by regional metamorphism. Ceylon contains sharply defined veins of graphite in gneisses interbedded with limestone. Graphite is also found: in the Adirondacks around Ticonderoga, New York, in the quartzites of the Grenville rocks; in Alabama in granites and schists; and in gneiss of the Pickering Valley, Chester County, Pennsylvania. A vast amount of carbon is held in the large and extensive coal, petroleum, natural gas and various petroleum waxes and other solids.
Uses: The largest single use of graphite is for facing sand molds for casting metals in foundries; it is also used in lubricants, batteries, and crucibles for melting metals. Other uses are in paints, "lead" pencils, electrodes for electric furnaces, electric reduction cells, and electroplating. Graphite can be synthesized in the electric furnace from coke, anthracite, or petroleum carbon.

Diamond *(illus. pp. 10, 75, 177)*

Isometric; hextetrahedral. The crystals are commonly octahedrons; dodecahedrons, tetrahedrons with modifying forms. Faces often strongly curved and striated; also in spherical forms. Twinning on {111} commonly producing contact spinel twins. It is composed of carbon; small amounts of other elements may be present, chiefly silicon, iron, aluminum, magnesium, calcium, and titanium, which are present as inclusions of graphite, magnetite, ilmenite, garnet, chromediopside, biotite, and possibly others. Colorless, yellowish to yellow, orange, reddish, rarely red, green, or blue white; often shades of brown or black due to inclusions which are probably very fine particles of carbon. L adamantine, often with a greasy appearance; H 10; SG 3.52; C {111} perfect but difficult to obtain; Fr conchoidal to uneven; brittle yet extremely difficult to break (otherwise it would not be usable in diamond drills); transparent to translucent, sometimes fluorescent.
Tests: Insoluble in acids and infusible. Burns brilliantly in an atmosphere of oxygen at 800° C; not attacked by alkalies. Distinguished by the hardness, adamantine to greasy luster, crystal forms, infusibility, insolubility.
Occurrence: Diamond originates in basic igneous rocks that con-

tain a high percent of magnesium and carbon in carbonates, graphite, and probably much carbon dioxide deep in the magma column in volcanic necks. The erosion of volcanoes and the core of diamond-bearing igneous rock liberated many diamonds, which were added to the heavy, resistant concentrates of gravel deposits of streams or along seacoasts. Such diamondiferous deposits may become consolidated again and weathered to release the diamonds and other resistant constituents of the original gravels.

Diamonds were found in Golconda and elsewhere in India from very early times; were discovered in Brazil in 1729 by prospectors searching for gold; and were in earlier times found in Borneo. The great diamond fields of South Africa were discovered in 1867 by children who found a few fine stones in the gravels of the Vaal River. The continued search revealed the presence of numerous diamonds in gravels and volcanic necks in South Africa, South-West Africa, Zaire, Ghana, and Tanzania.

The largest diamond was discovered in the Premier Mine in the Transvaal and weighed 3,106 carats. The other large diamonds vary in weight up to nearly 1,000 carats. There are no known stones between 1,000 and 3,000 carats.

Diamonds have been found in the loose soil and gravels at scattered localities in the United States—Virginia, North Carolina, Georgia, Wisconsin, and Illinois. In April, 1928, a well-formed dodecahedral diamond crystal of good color weighing 34.46 carats was found at Peterstown, West Virginia. It is believed to be the largest alluvial diamond found in the United States. Near Murfreesboro, Arkansas, diamonds were found in soil derived from weathered peridotite and also in the rock, which is similar to that of the diamond mines of South Africa. The largest Arkansas stone weighed 40.22 carats.

There are various types of diamonds, three of which are important:

Bort is a gray to black low-grade diamond. It is granular, without the perfect cleavage of the single crystal or larger grains, which is due to the large number of randomly oriented grains. The term "bort" is also applied to diamond crystals or grains of poor color or those that are badly flawed or contain impurities or inclusions. A major part of the diamond production falls into this category.

Ballas consists of rounded pieces with a radial granular structure that adds to their strength and toughness. It comes chiefly from Brazil and to some extent from South Africa.

Carbonado occurs as massive, compact, granular, dense aggregates of grains, black to gray black. Carbonados range in size from less than a carat to several thousand carats in weight. Due to the granular texture, they are less brittle than single crystals. They originate only in Bahia, Brazil.

Uses: The clear or colored transparent minerals are highly prized for gems (see Chapter 8).

The diamond is the superabrasive in many special industrial applications for which no other known substance can be adequately substituted. The industrial uses absorb about 80 percent of the diamonds produced, and beauty here has no particular value; consequently, the division between diamonds suitable for gems and those for industrial uses is sharply drawn, with a corresponding difference in value.

One of the first industrial applications of diamond—beyond the use of the dust in cutting and polishing diamonds and other hard gems—was that of the diamond core drill for exploring the extent and grade of ore deposits. The original bits contained a relatively few pieces of the tough carbonado from Brazil. Lacking an adequate supply for the increased demand, the present diamond core bits consist of a large number of small crystals or fragments of bort embedded in a metal matrix. When the diamonds are worn beyond the gauge tolerance, they are recovered by dissolving the metal with acid; they are then reset into other bits until they are too small, at which point they are crushed for use as powder or incorporated into bonded diamond abrasive wheels. The diamond core drill is used extensively for exploring the rock structure in connection with the building of dams, bridges, large buildings, aqueducts, and tunnels.

Modern industry requires many diamond-pointed tools for trueing and shaping grinding wheels and for grinding shaped tools of carbon and high-speed steels. Diamond-impregnated wheels are necessary for shaping the carbide tools used in lathes, shapers, and milling and boring machines.

Many tools tipped with formed diamonds are used in the finishing operations on precision-made machine parts in order to hold the required tolerances accurately over a long run.

In the stone-cutting industry, large saws up to eight feet in diameter having diamonds are used to cut building dimension blocks and monumental stones.

Diamond-pointed tools have a great variety of uses in many phases of our industry to such an extent that the lack of diamonds would have a decidedly adverse effect on production. The recent development of synthetic diamonds is providing a supply of smaller diamonds to the industry. It is the 80 percent of the natural diamond production that contributes so much to our well-being, rather than the one-fifth of the diamonds used for gems, which because of its esthetic appeal far overshadow the more useful part.

Rhodochrosite (Argentina), an attractive gem- and decorative stone; its rose color is due to manganese, a metal with a vast number of uses.

CERIUM (Ce) AND THE RARE EARTHS

The rare earths are a group of similar elements that cannot be easily separated by normal chemical methods. These elements, once thought scarce, are now known to be rather abundant in the earth's crust. The rare earths include cerium, lanthanum, praseodymium, neodymium, promethium, samarium, yttrium, europeum, gadolinium, terbium, dysprosium, holmium, erbium, thulium, ytterbium, and lutetium. These elements occur in more than 150 minerals, most of which are very rare and occur in small amounts in pegmatites. The minerals allanite, cerite, euxenite, fergusonite, formanite, monazite, and samarskite are important ores of the rare earth elements.

Cerium is a powerful deoxidizing agent. Cerium alloys with iron yield showers of sparks when struck or abraded, and are used in fireworks, cigarette lighters, sparking toys, and similar applications. Rare earth metals are principally used in petroleum refining, as cracking catalysts. They have specialized applications in explosives, photography, and glassmaking.

Allanite (Orthite)

Monoclinic, prismatic; tabular crystals parallel to {100} and acicular crystals parallel to the *b* axis; also massive, granular, or bladed aggregates and in disseminated grains. Named for T. Allan, who first described the mineral as a species. $(Ca,Ce,Fe,La)_2(Al,Fe)(SiO_4)_3$, composition very variable, may contain up to 20 percent rare-earth oxides; also uranium and thorium oxides. Black to brownish black to brown and sometimes coated with yellowish or brownish alteration products. L submetallic, pitchy to dull; S greenish gray to brown; H 5.5 to 6; SG 3.5 to 4.2; Fr conchoidal to uneven; brittle; translucent only on very thin edges. To check the presence of thorium or uranium, an autoradiograph of long exposure is needed. Allanite is a paramorph after an earlier mineral; the change involves a considerable increase in volume which causes radial fracturing in the surrounding rock minerals.

Tests: F 2.5 with intumescence to a black magnetic glass; CT some varieties give much water, gelatinizes in HCl. Gives reactions for rare earths. In the field the radial fracturing about the mineral shows it to be a paramorph after an earlier mineral with a smaller volume. The fracturing is not the result of radioactivity as some suppose. Magnetic glass BB separates it from uraninite. The mineral is difficult to distinguish from euxenite and polycrase without chemical tests.

Occurrence: A common accessory mineral of igneous rocks, especially the pegmatites, as in the anorthosites of Whiteface Mountain, New York; as large crystals in the magnetic iron ore deposits at Moriah, New York. Good crystals have come from Arendal, Norway; in acicular prismatic crystals up to a foot long from Finbo near Falum, Sweden; from Miask in the Ilmen Mountains, USSR. Fine crystals from Madawaska, Ontario. Also at Franklin, New Jersey; Chester County, Pennsylvania; Amelia County, Virginia; California; Massachusetts; Wyoming; and Barringer Hill pegmatite, Llano County, Texas, in association with gadolinite and other rare-earth minerals. Allanite is probably the most abundant and widely distributed rare-earth mineral and is a source of cerium and other rare earths.

Cerite

Orthorhombic, crystals rare and highly modified in short prisms; commonly massive and fine granular. A hydrous silicate of rare-earth metals, $(Ce,La,Y,Ca)_2Si(O,OH)_5$, with small amounts of iron, magne-

sium, and manganese. Gray, brown, cherry red; L dull adamantine; S grayish white; H 5.5; SG 4.86 to 4.91; fracture splintery; brittle; subtranslucent.
Tests: BB infusible; gelatinizes with HCl; CT yields water; borax bead yellow in oxidizing flame, almost colorless when cool; fuses to a dark, slaggy mass. The cherry red color is so characteristic that other properties hardly need testing. However, the relatively high SG and H just below feldspar are good supporting criteria. The test for rare-earth metals is confirmatory evidence.
Occurrence: In gneiss with allanite at Bastnäs, Sweden; in irregular lenses and stringers in aplite and pegmatite at Jamestown, Colorado; in the mountain pass district, California.
Use: An ore mineral of the cerium group of rare-earth metals, which may become of importance in the future.

A thermostat (l.), whose operation depends on mercury (in the glass vial). Cinnabar (r.) (California), the sole important ore of mercury.

Euxenite (Polycrase)

Orthorhombic, bipyramidal; tabular parallel to {010} or short prismatic parallel to the c axis, often in parallel or sub-parallel aggregates and in masses; commonly twinned on {201}; twins flattened and striated parallel to prism faces. The name is derived from the Greek *euxenos*, "hospitable," in allusion to the number of rare elements it contains. A titanate-niobate-tantalate of yttrium, calcium, cerium, thorium, and uranium, $(Y,Ca,Ce,U,Th)(Cb,Ta,Ti)_2O_6$. The composition is variable with some variations in the physical properties, especially the specific gravity. The high columbium-tantalum member of the series is euxenite proper, while the high titanium end of the isomorphous series is polycrase. Brown to black; L submetallic, brilliant on fresh surface, also inclined to be vitreous or greasy; S yellowish, grayish, or reddish brown; H 5.5 to 6.5; SG 4.9 to 5.9; Fr conchoidal to subconchoidal; brittle; translucent along thin edges. An amorphous paramorph after an earlier mineral crystallized at a high temperature. Distinguished with difficulty from eschynite-priorite, as the properties overlap and the compositions are similar.
Occurrence: Pegmatites and fine-grained granite dikes are the chief sources. Euxenite-polycrase minerals are like allanite, changing from a crystalline to an amorphous structure while retaining the crystal form of an earlier mineral; during the change, the volume increases, in such proportions as to produce a radical fracturing in the enclosing minerals. These minerals are widely distributed in numerous localities in Norway; Sweden; Finland; Madagascar; Zaire; Namaqualand, South Africa; and Brazil. In the United States polycrase is found in North and South Carolina; Barringer Hill, Llano County, Texas; and the Adirondack Mountains, New York. Euxenite occurs in Delaware County, Pennsylvania, and polycrase is found in the Nipissing District of Ontario.
Uses: These minerals are possible sources of the rare earths thorium, uranium, columbium (niobium), and tantalum.

Fergusonite, Formanite

Tetragonal, bipyramidal; crystals are prismatic or pyramidal, also as long prismatic parallel to the c axis; also in irregular masses and grains. Fergusonite named for Robert Ferguson, a Scottish physician; formanite named for Francis G. Forman, government geologist of western Australia. A columbate and tantalate of yttrium with erbium, cerium, lanthanum, and other rare earths and radioactive elements. Fergusonite $(Y,Er,Ce,Fe)(Cb,Ta,Ti)O_4$, formanite $(U,Zr,Th,Ca)(Ta,Cb,Ti)O_4$; fergusonite is the essentially pure columbium end member while formanite is the tantalum end member. Brownish black to

black, lighter color on alteration surfaces with gray, yellow, or brown color; L on fresh surface submetallic to brilliant vitreous; S brown, yellowish brown, greenish gray; H 5.5 to 6.5; SG 5.6 to 5.8, variable with composition; C {111} in traces; Fr subconchoidal; brittle; translucent only in the thinnest edges.
Tests: BB infusible; CT may yield some water on ignition; partly decomposed by H_2SO_4 and HCl. Completely decomposed by fusion with potassium bisulfate, $KHSO_4$; fused BB with borax and dissolved in HCl and boiled with tin yields a blue-colored solution, columbium (niobium).
Occurrence: These are pegmatite minerals. Fergusonite localities are numerous in Norway, Sweden, Finland, and the Ural Mountains; in placers in Swaziland, southern Rhodesia, Mozambique, and Madagascar. Formanite is common in western Australia. Fergusonite occurs in detrital materials from pegmatites in North and South Carolina; and in large crystals in Virginia and at Barringer Hill, Texas; also in Riverside and San Diego Counties, California; El Paso County, Colorado; Massachusetts; and Connecticut. These are ore minerals of the rare earths and radioactive elements.

Monazite

Monoclinic, prismatic; crystals usually small, sometimes flattened parallel to the front pinacoid or elongated parallel to the b axis; may be wedge-shaped or equant; commonly twinned on the base with occasional cruciform twins. Name derived from the Greek *monazein,* "to be solitary," in allusion to its isolated crystals. A phosphate of the cerium group rare-earth metals, $(Ce,La,Y,Th)PO_4$. The composition is variable and the amount of thorium may vary from a small percent up to 15 percent. Small amounts of calcium, magnesium, aluminum, and silica may be present. Yellowish or reddish brown to brown and sometimes white; L vitreous to subadamantine; S white or nearly white; H 5 to 5.5; SG 4.6 to 5.4; C {100} distinct; brittle.
Tests: BB infusible, decomposed by fusion with potassium bisulfate or sodium carbonate; slowly decomposed by hot acids; thorium-rich material is radioactive. The occurrence, color, and SG are the best identifying characters.
Occurrence: Monazite is an accessory mineral of granites, granitic metamorphic rocks, and pegmatites, and is widely distributed in detrital material from these rocks. Associated with the rare-earth minerals xenotime, gadolinite, samarskite, and fergusonite. The localities are numerous in the pegmatite areas of Norway, Sweden, Swaziland, western Australia, Brazil, and Ontario. In the United States monazite is found especially in New England and the Carolinas; at Climax, Colorado; Encampment, Wyoming; and Amelia, Virginia. Monazite is the chief ore mineral of the rare earths and thorium.

Samarskite

Orthorhombic, crystals prismatic, parallel to the c axis or tabular parallel to the front or side pinacoids; also massive and granular. Named for a Russian mine official, von Samarski. A columbate and tantalate of the rare-earth elements, calcium, iron, thorium, and uranium $(Y,Er,Ce,U,Fe,Ca,Th)(Cb,Ta)_2O_6$; small amounts of magnesium, manganese, lead, tin, and zirconium may be present. Brown to black; L vitreous to resinous on fresh surface; S dark reddish brown to black; H 5 to 6; SG 5.6 to 5.8; C {010} indistinct; Fr conchoidal; brittle; transparent on very thin edges.
Tests: BB fuses with difficulty to a black glass on edges; decomposes in hot acids; CT decrepitates, glows and turns black, decomposed by fusion with $KHSO_4$; when dissolved with HCl and boiled with metallic zinc gives a fine blue color. Similar to other rare-earth minerals, usually more radioactive than most of the other similar minerals.
Occurrence: Chiefly in pegmatites associated with columbite, monazite, cleavelandite, zircon, beryl, uraninite, muscovite, albite, quartz, topaz, garnet, and tourmaline. Found at numerous localities in Norway, Sweden, Madagascar, and Brazil. Large masses up to twenty pounds have been found in the pegmatites of North Carolina; also found in pegmatites in New England (especially Topsham, Maine), California (at Neuvo), near Idaho City, Idaho, and in Quebec.
Use: An ore mineral composed largely of useful elements but too rare to supply any significant quantity of the constituent elements.

CESIUM (Cs)

Cesium was the first metal discovered by Bunsen and Kirchoff in 1860 with the newly invented spectroscope. It is an essential constituent of only a few minerals, probably less than a dozen. Cesium is a white metal with a melting point of only 28.5° C and SG of 1.873. Cesium is the softest of all solid metals, and is also one of the most reactive. Its great affinity for oxygen makes cesium useful in removing traces of gases from vacuum tubes. Cesium is also used in electronic-eye devices, as a catalyst, and in cesium-vapor lamps. Only a few cesium minerals could be considered commercially important.

Pollucite

Isometric, rarely in small cubes, usually massive or in small grains, masses often show a lamination which is distinctive. Named for Pollux of Greek mythology. A hydrated alumi-

num silicate of cesium and sodium of variable composition, approximately $Cs_4Al_4Si_9O_{26} \cdot H_2O$. Colorless to white; L vitreous; S white; H 5.5; SG 2.90 to 2.98; C indistinct; Fr conchoidal; brittle; transparent.
Tests: BB fuses with difficulty turning white; imparts yellow color to flame (sodium). CT becomes opaque and yields water at high temperature. Decomposes in HCl with separation of silica. Gives characteristic lines in the orange and blue with the spectroscope. Distinguished from quartz, which it closely resembles, by its inferior hardness, as it is easily scratched by quartz or feldspar.
Occurrence: A mineral of the sodium-lithium pegmatites; occasionally in granite, as found on the island of Elba. Found in large masses at Karibib, South-West Africa. In the United States in pegmatites of Maine at Hebron, Greenwood, Newry, Mount Mica, and Buckfield; at Leominster and Goshen, Massachusetts; and in pegmatites in Custer County, of South Dakota, especially Tin Mountain Mine; also in Pala, California.

Pollucite is characteristically associated with cleavelandite, colorless and light-colored tourmaline, colorless beryl (goshenite), cesium beryl, spodumene, cassiterite, lepidolite, cookeite, and quartz. It is the most important ore mineral of cesium.

CHROMIUM (Cr)

Chromium, whose name refers to its brightly colored compounds, was discovered and isolated in 1707–1708. About twenty minerals contain chromium as an essential element, but there are many other minerals in which chromium substitutes for another element, including beryl, corundum, tremolite, diopside, and mica. Chromium is tremendously vital to the steel industry, for its role in making ferrochrome, an alloy with iron containing sixty to seventy percent chromium. Chrome steels, usually containing from one-fourth percent to three percent chromium, are noted for their hardness, strength, toughness, high ductility, and shock resistance. Stainless steel contains chromium, nickel, and iron; it is essentially nonmagnetic, has a pleasing white luster, and is widely used in jet engines, structural uses, appliances, sinks, labware, hotels, restaurants, hospitals, and in cutlery and holloware. Chromium metal is used as a protective coating in making auto bumpers; as a hard, wear-resistant surface for bearings; and as a decorative surface finish. Chromium compounds are used widely by the chemical industry for tanning leather, making dyes, oxidizing agents, pigments, and paints. Other compounds are used in photography, bleaching, surface treatment of metals, and as a cleaning agent.

Chromite *(illus. p. 131)*

Isometric, usually granular-massive; a chromate of ferrous iron, $FeCr_2O_4$; magnesium, aluminum, and ferric iron are sometimes present in significant amounts. Black; L metallic; S brown; H 5.5; SG 4.2 to 4.8; C none; brittle; transparent only in thinnest splinters or edges. May be feebly magnetic due to included magnetite.
Tests: BB essentially infusible; insoluble in acids; decomposed when fused with $KHSO_4$; borax bead and salt of phosphorus bead give reaction for iron when hot but give a fine green color (chromium) when cold. Chromite is similar to magnetite, spinel, and franklinite; all may be slightly magnetic if inclusions of magnetite are present. Chromite octahedrons may usually be identified by the lack of twinning and absence of strong magnetic properties. Chromite is softer and blacker than spinel. It alters to limonite and sometimes to stichtite.

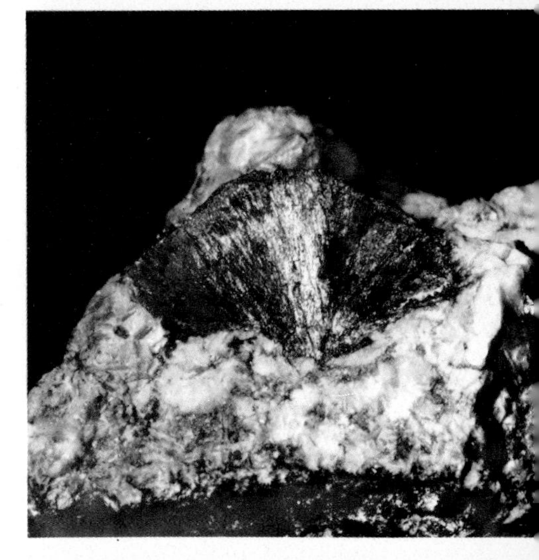

Top: Wulfenite from Red Cloud Mine, Ariz. *Center:* Millerite, from the Transvaal, an ore of nickel. *Above:* Canadian silver medal, surrounded by two silver pieces (Kongsberg, Norway; Canadian Northwest Territories) and a piece of silver-safflorite (center) (Cobalt, Ontario).

Occurrence: Chromite is a constituent of basic rocks such as peridotites, as disseminated accessory grains or segregations in the parent rock. The associated minerals are frequently spinel, idocrase, uvarovite, olivine, pyroxene, pyrrhotite, and niccolite. Among the important deposits are those in the Ural Mountains associated with platinum; Asia Minor; in the serpentine of the Great Dike of southern Rhodesia; Norway; Yugoslavia; the Philippine Islands; India; New Caledonia; and New South Wales. Only relatively small deposits occur in the United States: in the Coastal Ranges from Santa Barbara northward; in North Carolina, Oregon, Washington, and Wyoming; and in Lancaster County, Pennsylvania.
Uses: Chromite is the chief and practically the only ore mineral of chromium and its compounds. The mineral is a high-grade refractory and is used in making bricks; the ground material is mixed with varying amounts of other refractories, which are burnt with coal tar for refractory lining in open hearth and other furnaces.

Crocoite *(illus. p. 131)*
Monoclinic, prismatic; usually in prismatic crystals parallel to the c axis; sometimes pyramidal. Hollow crystals are frequent, also columnar, granular, and massive crystals, often highly modified and striated. Name derived from the Greek *krokos,* "saffron," in allusion to the color of the mineral. A chromate of lead, $PbCrO_4$. Yellow, orange, orange red to hyacinth red; L adamantine; S orange yellow; H 2.5 to 3; SG 5.97 to 6.2; C prismatic {100} distinct; Fr small conchoidal to uneven; brittle; sectile; transparent to translucent.
Tests: BB fuses easily; with sodium carbonate yields a lead button on charcoal; the lead volatilizes easily leaving a dark residue of chromium oxide; the lead gives a yellow coating

From near Potosí in Bolivia, fine crystals of cassiterite, which is the only important ore of the versatile metal tin.

around the assay. CT decrepitates, turns black while hot, but regains its orange red color when cold. Borax and salt of phosphorus beads yield grass green color when cold. Soluble in nitric acid, yielding a yellow solution. Its hyacinth red color and adamantine luster distinguish it from wulfenite, as does its lower SG and prismatic habit; distinguished from realgar by blowpipe reactions.
Occurrence: A secondary mineral in oxidized lead-chromium veins associated with pyromorphite, cerussite, vanadinite, and other secondary minerals. Found in the Ural Mountains, Romania, and Brazil. Tasmania has furnished fine crystals and crystal groups. Also from Darwin Mine, Inyo County, California, and Mammoth Mine in Arizona. A rare, highly prized mineral.
Uses: An ore mineral of lead and chromium but too rare to be of commercial importance. Crystal groups command good prices for cabinet specimens.

COBALT (Co) *(illus. p. 127)*
Cobalt is not an abundant element, occurring as an essential constituent in about twenty or thirty minerals, of which less than a dozen are important sources of the metal.

Cobalt melts at 1495° C.; SG 8.9. The metal is brittle, hard, and next to iron in magnetic properties. The properties and uses of cobalt are similar to those of chromium and nickel. The pure metal is used widely in electroplating, since its appearance and hardness are superior to nickel plate. Most of the cobalt used in the United States is in alloys. Three to four percent of cobalt in steel produces a high-speed metal-cutting alloy, which retains its hardness even at red heat. Cobalt steel is highly magnetic and is used for permanent magnets. Alloys such as alnico (aluminum, nickel, cobalt, and iron) are highly magnetic and will lift up to sixty times their own weight of iron. Stellite, a very hard alloy, is composed of chromium, tungsten, and cobalt, with small amounts of silicon, manganese, and carbon. This alloy is used for high-speed cutting tools, oil-well bits, scraper edges, shovels, and earth-moving equipment. Cobalt steels are used in gas turbines and turbo superchargers.

Cobaltite

Isometric; tetartohedral; usually in cubes or pyritohedrons or in combinations of these with the cube faces striated parallel to the axes; also in octahedrons and as granular compact masses, sometimes twinned on the dodecahedron or octahedron face. Named in allusion to its composition. Cobalt sulfarsenide, CoAsS. The mineral usually contains up to 5 percent iron and several percent nickel. It is isomorphous with gersdorffite, NiAsS. The color is white to light gray metallic with a faint reddish to violet tint. L metallic, bright on polished surface; S grayish black; H 5.5; SG 6.1 to 6.3; C cubic, perfect; Fr uneven; brittle; opaque; pyroelectric.
Tests: BB fuses readily to a weakly magnetic globule, gives a sublimate with an odor of garlic. In CT unaltered; OT gives odor of sulfur and an acid reaction on moist litmus paper. Also a sublimate of As_2O_3. The roasted material gives a blue color to the borax and salt of phosphorus beads. Decomposed by HNO_3 with the separation of sulfur.
Occurrence: Occurs in intermediate- and high-temperature veins, usually associated with smaltite, safflorite, gersdorffite, and copper and zinc minerals. Found at Lunaberg, Sweden, in pyritohedrons; in Silesia, Norway, and elsewhere. In fine octahedral crystals at Cobalt, Ontario. Also at Copper King Mine, Boulder City, Colorado, and in Lemhi, Idaho. An ore mineral of cobalt and of arsenic as a by-product.

Erythrite

Monoclinic, prismatic; in acicular prismatic crystals striated deeply parallel to the c axis; good crystals rare, usually as rounded globular reniform crusts with a drusy surface; earthy masses and powdery. Name derived from the Greek *erythros,* "red," in allusion to its unusual color. A hydrated arsenate of cobalt, $Co_3(AsO_4)_2 \cdot 8H_2O$; may contain varying amounts of annabergite. Erythrite is crimson but becomes paler rose red with increased nickel; L vitreous, also dull and earthy; S paler than the massive material; H 1.5 to 2.5; SG 3 to 3.15; C {010} perfect; translucent, transparent.
Tests: Fuses easily to a gray bead imparting a bluish color to the flame (arsenic) and gives a white sublimate; after roasting gives a deep blue bead with borax. In CT produces much water, and when mixed with charcoal and heated intensely gives an arsenic mirror; in OT yields a white sublimate. Soluble in HCl, yielding a rose-colored solution. Distinguished by its color and blow pipe tests.
Occurrence: Erythrite is a secondary mineral usually resulting from the oxidation of cobaltite, smaltite, and other cobalt minerals. Fine specimens have been found in Germany, Austria, Switzerland, Sweden, and England. Beautiful crystal specimens come from Howden, South Australia. In the United States it is found in Nevada, Arizona, New Mexico, California, and the Blackbird district, Lemhi County, Idaho. In Canada it is found abundantly in Cobalt, Ontario.

Erythrite is an excellent guide to cobalt-nickel-silver ores with which it is usually associated. Every prospector should be familiar with it. It was the great showing of earthy ery-

Rutile (Graves Mountain, Georgia), an ore of titanium—a lightweight metal used in airplane construction and in the making of white pigment.

thrite, known as "cobalt-bloom," on the rocks at the eastern end of Great Bear Lake, Canada, that led to the discovery of the enormously valuable deposit of silver, uranium, and cobalt—one of the great romances of prospecting and mining.
Use: An ore mineral of cobalt.

COLUMBIUM (Cb), TANTALUM (Ta)

Columbium (niobium) and tantalum are rare metals with important technological uses today. Pure columbium was not isolated until 1906, but for more than twenty years only a small piece of it existed. Tantalum is a white, silvery metal with an extremely high melting point, 2996° C. Tantalum is ductile and malleable, especially when hot, but with a very high tensile strength. It is especially valuable because of its resistance to acids. Columbium is an important stabilizing metal for certain types of stainless steel.

Minerals containing niobium and tantalum occur chiefly in igneous rocks, especially pegmatites. Many of these minerals are chemically very complex.

Columbite-tantalite

Columbite is orthorhombic, prismatic; crystals are usually tabular and sometimes very large; also granular, massive. A columbate of iron and manganese, $(Fe,Mn)Cb_2O_6$. Color iron black to reddish brown; L brilliant submetallic; S reddish brown to black; H 6; SG 5.2 to 6.3, Fr subconchoidal; brittle. Tantalite is an iron manganese tantalate, $(Fe,Mn)Ta_2O_6$. Black to brownish black; S brownish black; H 6 to 6.5; SG 18.2; {010} distinct; Fr uneven to subconchoidal; brittle.
Tests: BB infusible; when fused with borax, dissolved in HCl, and boiled with tin gives a blue solution. Tantalite gives only a slight reaction for columbium. Borax bead in OF is green (manganese). F on charcoal with a little Na_2CO_3 yields a magnetic mass (iron). There are no simple tests for the detection of tantalum.
Occurrence: Both minerals are found in coarse-grained acid igneous rocks with microcline, quartz, spodumene, muscovite, lepidolite, and many other pegmatite minerals. From many localities in the Black Hills of South Dakota, which have yielded good crystals up to 200 pounds and masses up to 18 tons. Also from many localities in New England, especially Stoneham, Maine; Acworth, New Hampshire; Chesterfield, Massachusetts; and Haddam, Middletown, and Portland, Connecticut. Other localities are Canada, Brazil, Norway, Italy, Zaire, South-West Africa, and Madagascar. Ore minerals of columbium and tantalum.

COPPER (Cu) *(illus. pp. 33, 115, 129, 165, 168)*

Copper has been known and used by man since before the dawn of recorded history, possibly as long as 20,000 years ago. In the United States the extensive native copper deposits of the Great Lakes area, especially Michigan, were known and worked by the American Indians or their predecessors. The uses of copper in our present society are so widespread and vital that a volume would be required to enumerate them. Copper is the backbone of the electric power industry, for generation and conduction of electricity. It is present in nearly all electrical devices, appliances, computers, transportation systems, and factories, homes, and businesses. Copper is present in many alloys, such as brass, that are also extensively used.

There are more than 150 minerals containing copper as an essential element. About thirty of these may be considered ore minerals. But to the collector of mineral specimens, copper minerals are in a class by themselves, because most of them are brightly colored in shades of red, brown, green, and blue.

Azurite *(illus. pp. 25, 40, 129)*

Monoclinic, prismatic; crystals are varied in habit, often highly modified; tabular, short prismatic; equant; striated; in composite or parallel groupings, also in spherical groups and rosettes, stalactitic, coarse radial structure as well as dull and earthy. A basic carbonate of copper, $Cu_3(OH)_2(CO_3)_2$, containing 55 percent copper. Color varies from light to dark blue; L vitreous; S light blue; H 3.5 to 4; SG 3.6 to 3.8; C {011} perfect; Fr conchoidal; brittle; transparent to translucent.
Tests: BB turns black, fuses readily yielding a copper globule; soluble in dilute acid with effervescence; gives reaction for copper; CT blackens, yields water; soluble in ammonia yielding a deep blue solution. Distinguished by blowpipe tests.
Occurrence: A secondary mineral found in the weathered parts of copper deposits. Often associated with malachite in concentric bands. Alters to malachite; associated with cuprite, tenorite, chalcocite, chrysocolla, calcite, limonite, wad, and other alteration products of primary minerals. Azurite occurs in many copper mines, such as those of the Ural Mountains; the Balkan countries; Chessy, France; Tsumeb, South-West Africa; common in the copper mines of the western United States—at Bisbee and Morenci, Arizona, for example—and other copper-producing countries.
Uses: An ore mineral of copper. Once used by artists for a blue pigment in paint; used to a minor extent for decorative purposes; a semiprecious stone. Fine azurites make prized cabinet specimens.

Bournonite

Orthorhombic, bipyramidal; short prismatic and tabular with striated faces; granular; compact; massive. Twinned on

prism {110}, often cyclic, producing cruciform or notched circular aggregates referred to as "cogwheel" ore by miners. Named for Count J. L. de Bournon, French crystallographer and mineralogist. A sulfide of lead, copper, and antimony, $PbCuSbS_3$; arsenic may be present occasionally. Gray; L metallic, brilliant to dull; S gray to dark gray; H 2.5 to 3; SG 5.8 to 5.85; C {101} imperfect; Fr uneven to subconchoidal; brittle; opaque.
Tests: BB on charcoal fuses easily to a metallic globule yielding a sublimate of antimony. Decomposed in HNO_3, yielding a blue solution which will give a copper plate on iron. Distinguished by twinning or blowpipe tests.
Occurrence: Bournonite is a common mineral associated with many other minerals in the intermediate temperature veins. Some important localities are in the Harz Mountains, Germany; Pribram, Bohemia; Kapnik and elsewhere in Romania; Cornwall, England; also in Spain, Italy, and France. In the silver-tin mines of Bolivia, also in Chile and Peru; at Broken Hill and elsewhere in Australia. In the United States, in Arizona, California, Nevada, Colorado, Montana, Arkansas, and in large crystals at Park City, Utah. An important ore mineral of copper; also of lead and antimony.

Chalcopyrite *(illus. pp. 34, 165)*

Tetragonal, scalenohedral; occasionally in small bisphenoids, sometimes modified by scalenohedrons which are striated; usually granular, massive, compact; sometimes rounded masses, botryoidal, or reniform. The chalcopyrite scalenohedrons often resemble the tetrahedrons of the isometric system. A sulfide of copper and iron, $CuFeS_2$; in gold-bearing areas, small to appreciable amounts of gold are associated with the mineral. Metallic, light orange yellow; L bright metallic to dull, often tarnished and iridescent; S greenish black; H 3.5 to 4; SG 4.1 to 4.3; C {011} usually indistinct; Fr uneven to subconchoidal; brittle; opaque.
Tests: BB on charcoal decrepitates, fuses readily to a magnetic globule in the RF; with HNO_3 gives a white sulfur residue, solution gives tests for iron and copper. Distinguished by small bisphenoidal crystals, which are characteristic; softer than pyrite; resembles gold, but is brittle and gives a black powder.
Occurrence: It is the most widely distributed of the copper minerals and is associated with many metallic and nonmetallic minerals in intermediate and basic rocks. Many minerals are associated with chalcopyrite. The localities are legion in Europe and southern Africa and along the Andes of South America and the Rocky Mountains of North America. In the United States excellent crystals come from Chester County, Pennsylvania, and St. Lawrence and Ulster counties, New York. Near Joplin, Missouri, beautiful small sphenoids are abundantly sprinkled over dolomite and sphalerite. Widespread in copper deposits in Arizona, New Mexico, Utah, Montana, and Tennessee. Abundant in Rouyn and Sudbury districts in Quebec and Ontario; also in British Columbia and Manitoba. An important ore mineral of copper.

Chrysocolla *(illus. pp. 132, 192)*

Probably orthorhombic, in microscopic acicular crystals from Mackay, Idaho; usually amorphous or cryptocrystalline; massive; compact; botryoidal. Hydrated copper metasilicate, $CuSiO_3 \cdot 2H_2O$, containing 36.2 percent copper; often quite impure, commonly siliceous. Bluish green or greenish blue, also brown or black with impurities; L vitreous to earthy; S white; H 2 to 4; SG 2 to 2.4; C none; Fr conchoidal; extremely brittle; translucent to opaque.
Tests: BB infusible, decrepitates, colors the flame green; CT blackens and gives off water; gives copper reaction with the fluxes; with Na_2CO_3 gives copper globule on charcoal; soluble in acids without gelatinization. Distinguished from turquoise by brittleness and greater softness.
Occurrence: A common secondary mineral in the upper portions of copper deposits usually associated with azurite, cuprite, quartz, and calcite. Near Globe, Arizona, with a coating of drusy quartz making specimens of matchless beauty. Found in many copper mines all over the world. Notable material is found at Bisbee and Morenci, Arizona; Tintic district, Utah; Mackay, Idaho; also in Pennsylvania, Michigan, New Mexico, and at Tsumeb, South-West Africa; Katanga, Zaire; Cornwall, England; Schneeberg, Germany; and Adelaide, South Australia.
Uses: A minor ore mineral of copper; when intimately mixed with chalcedony it is cut into semiprecious stones. It is erroneously called blue chrysoprase.

Cuprite *(illus. p. 30)*

Isometric, gyroidal, octahedral, sometimes as cubes, less frequently as dodecahedrons; occasionally in a matted network of extremely elongated acicular or capillary crystals parallel to an a axis, giving a form known as chalcotrichite; also granular, massive, and earthy. Cuprous oxide, Cu_2O. Color varies from a red to brown or grayish red; L adamantine to submetallic in blackish specimens; S several shades of brownish red; H 3.5 to 4; SG 6.14; C {111} interrupted; Fr conchoidal to uneven; brittle; transparent.
Tests: BB on charcoal, blackens, fuses, and is reduced to metallic copper; unaltered in CT; in forceps, fuses and colors flame green; soluble in HCl, giving a blue solution; the concentrated cold HCl solution when diluted with water yields a heavy white precipitate of cuprous chloride,

CuCl. The crystal form together with the red adamantine luster and softness separate it from all other minerals. Differs from cinnabar and proustite by the color of the streak. Gives reaction for copper; others do not. Cuprite is the only red copper mineral.
Occurrence: Cuprite is common and of widespread occurrence in the oxidized parts of copper deposits; association with azurite, malachite, native copper, chrysocolla, and limonite. Octahedral and dodecahedral crystals from Chessy, France. Also occurs at Cornwall, England; Tsumeb, South-West Africa; in USSR, Germany, Hungary, and elsewhere in Europe; in Chile, Bolivia, and Mexico. Bisbee, Arizona, is famous for its fine crystals. Magnificent specimens of the variety chalcotrichite in long, slender cubic crystals come from Morenci, Arizona, and other copper mines. Other United States localities are Pennsylvania, Tennessee, Montana, and Colorado. An important ore mineral of copper.

Dioptase *(illus. p. 194)*

Hexagonal, rhombohedral, commonly in equant crystals, a combination of the prism and rhombohedron, also in prismatic crystals up to two inches long. Hydrous orthosilicate of copper, H_2CuSiO_4, containing 40.2 percent copper. Bright dark green; L vitreous; S green; H 5; SG 3.28 to 3.35; C perfect rhombohedral; Fr conchoidal; brittle; transparent to translucent.
Tests: BB infusible, blackens, decrepitates, colors flame green; reaction for copper with fluxes; on charcoal with fluxes gives a button of copper; soluble in HCl with separation of silica; CT blackens and yields water. Dioptase is one of the showiest of all minerals; its fine green color and excellent crystals will usually identify it at sight; blowpipe tests are conclusive; malachite dissolves in HCl with effervescence.
Occurrence: Dioptase occurs in the upper zone of weathering of copper deposits in dry or arid climates, such as the Ural Mountains and North Africa. Small but fine crystals were found in Arizona at several mines, notably Mammoth–St. Anthony Mine in Tiger. Also found at Soda Lake Mountain, San Bernardino, California.
Use: An ore mineral of copper. Crystal groups have good specimen value. Few minerals add as much vividness and attractiveness to a collection as dioptase.

Enargite

Orthorhombic, pyramidal, prismatic parallel to the c axis and tabular parallel to the base; also granular, massive, compact, columnar, and bladed. Crystals are usually vertically striated and small. A copper arsenic-antimony sulfide, $Cu_3(As,Sb)S_4$; iron and antimony may be present up to 5 percent. Grayish black to black; L metallic; S grayish black; H 3; SG 4.4 to 4.5; C {110} perfect; Fr uneven; brittle; opaque.
Tests: On charcoal fuses readily, yielding a faint white coating and garliclike odor; CT decrepitates; gives a sublimate of arsenic sulfide when roasted; when mixed with Na_2CO_3 and heated in RF gives copper button. Differs from other copper sulfarsenides by its perfect cleavage, from stibnite in that it cannot be bent; associated with rare copper arsenates and sometimes with azurite and malachite.
Occurrence: Enargite occurs chiefly in intermediate-temperature veins and other mineral deposits; found in Austria, South-West Africa, Peru, Chile,

and Mexico. In the United States in Utah, Colorado, California, Nevada, and Butte, Montana. An ore mineral of copper and arsenic.

Malachite *(illus. pp. 36, 40, 192)*

Monoclinic; rarely in short to long, prismatic crystals; commonly as crusts with delicate to striking banding, fibrous aggregates; compact; stalactitic; occasionally altering to azurite. A basic carbonate of copper, $Cu_2(OH)_2CO_3$. Light to dark green; L subadamantine, sometimes velvety; S pale green; H 3.5 to 4; SG 4.03 to 4.07; C {201} perfect; Fr subconchoidal; brittle; translucent to opaque.
Tests: BB blackens, fuses easily, coloring the flame a bright green; on charcoal reduces to a metallic button; soluble in acids with effervescence, green color, reactions for copper. Green chromium and nickel minerals do not effervesce.
Occurrence: In the upper parts of copper deposits; often surrounds copper exposed to the atmosphere as in roofs and telephone wires. Associated minerals are azurite and other copper minerals. Found in large amounts in the Ural Mountains; also in England; France; Germany; Romania; northern Zambia; southern Zaire; South-West Africa; abundantly at Bisbee, Arizona, in Grant and Socorro Counties, New Mexico, and elsewhere in the western United States.
Uses: An ore mineral of copper; compact varieties used for decorative material and semiprecious stones. An early use was for a green pigment.

Tetrahedrite-tennantite

An isomorphous pair; isometric-tetrahedral. The crystals usually have a pronounced tetrahedral habit; in groups of parallel crystals and usually in granular, compact masses. Twinned on {111} as contact and penetration twins; often cyclic twins are produced. This isomorphous pair can be well represented by the formula $(Cu,Fe,Zn,Pb,Ag)_{12}(Sb,As)_4S_{13}$, in which copper is the most abundant—approximately half—and the remaining elements variable. Sometimes enough silver is present to make a valuable silver ore. Medium gray to grayish black; L metallic; S red to brown to black; H 3 to 4.5; SG 4.6 to 5.1; C none; Fr uneven; brittle; translucent; transmits red light.
Tests: BB on charcoal fuses very easily; decomposed by HNO_3 with separation of sulfur and Sb_2O_3 in the varieties containing antimony. Solution becomes blue when ammonia in excess is added. Silver and lead varieties give a white precipitate when a chloride is added to the nitrate solution; the lead precipitate will dissolve upon heating. In crystals it is readily recognized by its tetrahedral crystals. A blowpipe check against the properties of similar minerals is usually needed.
Occurrence: In the intermediate to low-temperature veins. These minerals are widely distributed, tetrahedrite being the more common. Among the associated minerals are those of lead, zinc, copper, silver, iron, and barium. Important localities are Germany, Romania, Bohemia, England, Norway, Sweden, and Peru. In the United States these minerals are found in Colorado, Idaho, Nevada, New Mexico, California, Virginia, and Bingham, Utah. Fine crystals of tennantite occur in Cornwall, England; Butte, Montana; and Bolivia.

Opposite: Fine crystals of torbernite, a minor ore of uranium, from Shinkolobwe, Zaire. *Above left:* Vanadinite, from Apache Mine, Arizona, a desirable collector's mineral. *Above right:* Calcite (light) and zincite (dark), found almost exclusively in Franklin, New Jersey.

Uses: Important ore minerals of copper and at some places of silver and one or more of the other elements.

Turquoise *(illus. pp. 133, 188)*

Triclinic, pinacoidal; crystals are very rare, short prismatic, usually very small, dense to cryptocrystalline, compact as nodules. A hydrated basic phosphate of copper and aluminum, $CuAl_6(OH)_8(PO_4)_4 \cdot 4H_2O$. Turquoise is isomorphous with chalcosiderite, $CuFe_6(OH)_8(PO_4)_4 \cdot 4H_2O$. Bright intense blue to bluish green and nearly colorless. L waxy; S white or greenish; H 5 to 6; SG 2.6 to 2.83; subtranslucent to opaque.
Tests: BB infusible, turns brown and soluble in acid after ignition, otherwise insoluble. CT decrepitates, yields water and turns brown or black; colors the flame green; soluble in HCl. Distinguished by its greenish blue compact nodules and blowpipe reactions.
Occurrence: Turquoise is a secondary mineral found in veins in altered aluminous rocks and in the oxidized upper parts of the large disseminated copper deposits. Some localities are Lynch Station, Campbell County, Virginia; Los Cerillos, New Mexico; Arizona; Colorado; Nevada; and California.
Uses: A minor ore mineral of copper and a highly prized semiprecious gemstone.

GOLD (Au) *(illus. pp. 25, 181)*

The love of gold and its use by mankind started long before the event could be recorded, perhaps before there was spoken or written language, tens of thousands of years ago. Throughout history gold has had two major uses: as money and as decoration. Today gold has many technological applications as well.

Gold

Isometric, hexoctahedral; crystals rare, octahedral or dodecahedral; in leaves or plates in open places in quartz veins. Native gold often contains varying amounts of silver, sometimes small amounts of copper up to twenty percent; also palladium up to ten percent and rhodium up to forty-three percent. Light to intense yellow; H 2.5 to 3; SG 15.6 to 19.3; C none; Fr hackly on torn edges; opaque to all but green light for thicknesses up to one millionth of an inch. Malleable, ductile, sectile, good conductor of electricity and heat.
Tests: Fuses easily; not acted upon by fluxes or a single acid; if pure, gold is completely soluble in aqua regia. Distinguished by its yellow, metallic luster, high malleability, and SG. All gold minerals yield a button of gold BB on charcoal. All other yellow minerals are harder, and gold is the only one to leave a yellow, metallic streak.
Occurrence: Gold in quantity is rare; it is widely distributed in traces in acid and intermediate mineral deposits, especially those in quartz veins. Gold is practically always associated with quartz, pyrite, or chalcopyrite in veins of hydrothermal origin, which may be shallow or deep-seated. The most extensive deposits are in quartz conglomerates.

The associated minerals are pyrite—the most common and of the greatest importance because it often carries gold, sometimes in quantities—chalcopyrite, galena, sphalerite, tellurides, native bismuth, arsenic, stibnite, and cinnabar. Quartz-bearing veins in which any of these minerals occur should never be passed by without having representative samples assayed for gold. The history of the discoveries and mining of gold fill thousands of pages of fascinating accounts.

The most striking occurrences of gold are usually in the rich placers where the gold has occasionally accumulated in favorable potholes or other traps to the extent that little else is present. When such sites were discovered, the miners literally shoveled gold from the streams. In general, however, most placer operations survive on less than a dollar per cubic yard of material worked. At the old gold prices of $20.67 and $35.00 per troy ounce, most gold mines operated on ore from about $5.00 to $20.00 per ton of ore mined. The present high price of gold should rejuvenate many old, deserted properties.

The telluride-producing areas are mostly in Transylvania, Australia, California, Colorado, and Ontario. The largest producing gold area is in the vicinity of Johannesburg, South Africa. Nevada, South Dakota, and Utah are the principal producers in the United States.

IRON (Fe)

Iron may have been known and used in ancient Egypt as much as 7,000 years ago. The first iron used by man probably originated in meteorites that fell to earth. Smelting dates back to the fifteenth century B.C. The Iron Age began when carburizing and heat-treating were discovered as methods for hardening and tempering iron tools and weapons. Today's uses of iron are numerous and well known.

There are hundreds of iron-bearing minerals, but only a few have significance to mineral collectors. It should be remembered, however, that most rocks contain iron in some amount, and iron oxides and hydroxides are the red and brown colorations so commonly seen on weathered rock surfaces.

Goethite

Orthorhombic, dipyramidal; prismatic, striated, capillary to acicular grains in radial aggregates; also botryoidal, reniform, and stalactitic masses with a radial structure. Named for Goethe, the German poet, who was an ardent mineral collector and amateur mineralogist. Hydrogen iron oxide, $HFeO_2$, containing 68.3 percent iron. Yel-

low to yellowish brown to yellowish black; L submetallic; S yellow; H 5 to 5.5; SG 3.3 to 4.3; C {101} perfect; Fr uneven; brittle; translucent to opaque.
Tests: BB infusible; CT yields water; RF becomes magnetic; soluble in HCl and gives tests for iron. Radial texture separates it from limonite; color and streak distinguish it from hematite and magnetite.
Occurrence: As a weathering product; in the tropics forms widespread voluminous deposits of laterite; gossans are often composed in large part of goethite; principal ore mineral of Alsace-Lorraine. Also central USSR; southern Appalachians and areas in eastern Texas; in fine crystals at Pribram, Bohemia. The best localities are at Marquette and Negaunee, Michigan.

Hematite *(illus. pp. 61, 133)*

Hexagonal, scalenohedral; crystals are usually thin to tabular, sometimes form excellent rosettes, often occur as micaceous specular hematite schists; also botryoidal, stalactitic, reniform, and mammillary structures; in great earthy masses, granular to compact; in extensive beds of oölitic and other concretionary forms. Ferric oxide, Fe_2O_3, having 69.9 percent iron; titanium is often present in small amounts. Crystals shiny black. Earthy and compact varieties a dull red; S red; H 5 to 6; SG 5.75 for crystals; C none; Fr subconchoidal; brittle.
Tests: RF becomes magnetic; soluble in HCl and gives strong test for iron with potassium ferrocyanide. Distinguished by black crystals and the red streak; red varieties by color and streak, which separate hematite from other iron minerals; from cinnabar and cuprite by their markedly different blowpipe reactions.
Occurrence: An alteration of earlier iron minerals. Large deposits occur in the Lake Superior region; Steep Rock Lake, Canada; the Clinton iron ores extending down the Appalachian Mountains from New York to Georgia; Iron Mountain, Missouri; Knob Lake in Quebec-Labrador; outcrops on Bell Island, Newfoundland. Large deposits occur in Brazil and Venezuela. The most important ore mineral of iron.

Limonite *(illus. pp. 17, 90)*

Amorphous; an earthy mass of alteration products of iron-bearing composition, $Fe_2O_3 \cdot nH_2O$, with colloidally precipitated silica and clay minerals, manganese oxides, and other alteration products. Various intensities of brown; L dull; S yellowish brown; H 2 to 5; SG 2.7 to 4.3; C none; Fr subconchoidal to even.
Tests: RF becomes magnetic; soluble in HCl and yields a strong test for iron; CT gives water and changes to hematite plus impurities. Distinguished from hematite and magnetite by yellow streak and yielding water in CT.
Occurrence: Limonite is often formed as pseudomorphs after other iron minerals, especially limonite after pyrite, such as that from Lancaster, Pennsylvania, and Pelican Point, Utah. Also found in Salisbury and Kent, Connecticut; Fruitville, Pennsylvania; Virginia; Georgia; Tennessee; and Alabama.
Uses: Limonite and associated iron minerals are important sources of iron at numerous localities. Limonite furnishes a number of known pigments which are marketed as ochers and umbers.

Magnetite *(illus. pp. 57, 165)*

Isometric, hexoctahedral; usually in octahedrons, sometimes dodecahedrons, rarely cubes; granular, massive. Often as single euhedral crystals in chlorite schists. Ferrous metaferrite, $FeFe_2O_4$, containing 72.4 percent iron; the richest of all important iron minerals. Grayish black to black; L metallic; S grayish black; H 5.5 to 6.5; SG 5.17 to 5.18; C none; Fr uneven; brittle; opaque; strongly magnetic; occasionally naturally polarized. Known as lodestone, which attracts magnetic materials; has the unique habit of developing a granular arrangement of distorted octahedral crystals solidly arranged within massive magnetite.
Tests: BB nearly infusible; strongly magnetic; gives tests for iron; soluble in HCl; OF changes to hematite, which is nonmagnetic. Distinguished from all other black minerals by its strong magnetic properties.
Occurrence: Magnetite is one of the most abundant and widespread minerals; it occurs disseminated through many metamorphosed rocks—schists, gneisses, coarse-grained marbles—in igneous rocks and as detrital grains in placers. It is probable that some of the magnetite deposits in gneisses as in the Adirondacks may be metamorphosed sedimentary iron ores older than the Grenville series. Among the noteworthy deposits are those of Kiruna in the Scandinavian peninsula; Magnitogorsk, USSR; India; Liberia; South Africa; Mexico; Magnet Cove, Arkansas; Port Henry, Mineville, and Brewster, New York; Hamburg and Franklin, New Jersey; and French Creek mines in Chester County, Pennsylvania. An important ore mineral of iron.

Siderite *(illus. p. 133)*

Hexagonal, scalenohedral; crystals usually simple rhombohedrons; sometimes in rounded scalenohedrons; spherical forms known as spherosiderite are not uncommon; frequently in cleavable masses. Ferrous carbonate, $FeCO_3$, containing 48.3 percent iron. Siderite is completely isomorphous with rhodochrosite, magnesite, and partially with calcite. Shades of brown; L vitreous, pearly on cleavage planes; S white; H 3.5 to 4.5; SG 3.95 to 3.97; C $\{10\bar{1}1\}$; Fr uneven; brittle; transparent to translucent.
Tests: BB on charcoal, darkens

and becomes magnetic. F is difficult; dissolves in hot HCl with effervescence; solution reacts for ferrous iron with potassium ferricyanide; CT decrepitates and emits CO_2. Distinguished by brown color and crystal form; BB tests separate it from other carbonates. Intermediate members of the isomorphous series cannot be separated.
Occurrence: Frequently primary gangue mineral in metalliferous veins, occasionally the predominant mineral; most commonly occurs in large deposits in sediments; such deposits are worldwide and were the source of iron ore for many local forges during the early history of the United States. Siderite has been worked for iron in Germany, Austria, and the Balkan countries. Siderite occurs in Bolivia and Minas Gerais, Brazil; a large vein has been worked at Mine Hill, Roxbury, Connecticut. Excellent crystals were obtained from the tin mines of Cornwall, England; also from the Piedmont, Italy, and from France and Switzerland. The deposits at Ivigtut, Greenland, contain numerous well-developed crystals in the cryolite. Also found in Gap Mine, Lancaster County, Pennsylvania, and the Gilman district of Eagle County, Colorado.
Use: Siderite is an ore mineral of iron even though the iron content is low, because the ore is readily and economically smelted.

Vivianite

Monoclinic, prismatic; commonly prismatic, tabular, reniform, in stellate groups and in earthy masses. A hydrated ferrous phosphate, $Fe_3(PO_4)_2 \cdot 8H_2O$, containing 33.39 percent iron; some manganese, magnesium, and calcium are sometimes present. Colorless and transparent, becoming blue to bluish black on prolonged exposure to light; C {010} perfect; S white to dark blue; H 1.5 to 2; SG 2.67 to 2.69.
Tests: BB yields black metallic globule. Easily soluble in acids. Characterized by its bright blue color after long exposure to light and its micaceous habit.
Occurrence: A late mineral in cavities in ore veins as a secondary mineral, widespread often as concretionary shapes in clay. Fine crystals four inches long have come from the tin mines of Cornwall, England; also fine specimens from Leadville, Colorado. Occurs at many places in Germany, France, and USSR; also in Allentown, New Jersey, and Middletown, Delaware.
Use: No importance as an ore mineral. Good quality specimens have some value among collectors.

Potassium, an alkali metal found in feldspars and micas, pictured here in violent reaction with water. In compound form it has diverse uses —as a disinfectant, an antiseptic, and a fertilizer, among others.

LEAD (Pb) *(illus. pp. 127, 130, 137)*

Lead has been known and used for thousands of years. It is 50 percent heavier than iron (SG 11.35), giving rise to such expressions as "heavy as lead." The low melting point, 327.4° C, gives lead utility in special alloys. Lead is chemically very inert, a feature that makes it useful in pipes. The Romans constructed extensive pipelines of lead centuries ago. However, many lead compounds are poisonous, and it has been speculated that massive and widespread lead poisoning may have contributed to the fall of the Roman Empire.

Lead sheet is used as X-ray shielding and in storage batteries that supply electrical energy for many kinds of vehicles. Lead prevents engine knocking when added to gasoline. Lead alloys are used in making type, solders, and weights of various kinds. Lead is also used to shield cables and in modern pipelines.

Lead minerals are popular among collectors, as they form excellent and beautiful crystals.

Anglesite

Orthorhombic, dipyramidal, crystals often well developed, the habit varying from prismatic to tabular and equant. Usually granular, compact, massive; galena crystals are often coated or crusted with anglesite when exposed to weathering. Lead sulfate, $PbSO_4$, containing 68.3 percent lead; anglesite alters to cerussite. White to grayish white, yellow, green, blue; L adamantine to vitreous; S white; H 2.5 to 3; SG 6.4 to 6.5; C {001} good; Fr conchoidal; brittle; transparent to opaque.

Tests: F 1.5; BB on charcoal yields a metallic button in RF; soluble with difficulty in HNO_3; soluble in ammonium oxalate; gives tests for lead. It is frequently associated with galena; distinguished from barite by the perfect prismatic cleavage of barite; cerussite effervesces in nitric acid; anglesite is the heaviest of the light-colored, nonmetallic minerals.
Occurrence: A common weathering product of galena in upper portions of lead mines. Fine crystals from Scotland, Germany, South-West Africa, Australia, and Mexico. From Wheatley Mines, Phoenixville, Pennsylvania; Missouri lead mines; Eureka, Utah; Castle Dome district, Yuma County, Arizona; and Idaho. An ore mineral of lead.

Cerussite *(illus. p. 134)*

Orthorhombic, bipyramidal; crystals tabular, pseudohexagonal, equant; crystal groups are sometimes stellate, usually in reticulated aggregates, also granular, compact, earthy. Twinning is almost universal. Lead carbonate, $PbCO_3$, containing 77.5 percent lead; may contain small amounts of strontium, zinc, and silver if present in the primary mineral. White to variously colored depending on the impurities present; L adamantine to vitreous; S white; H 3 to 3.5; SG 6.5 to 6.6; C {110} and {021} distinct; Fr conchoidal; very brittle.
Tests: BB fuses easily, yields a metallic button in RF; gives tests for lead; a yellow coating on charcoal about assay; soluble in dilute HNO_3 with effervescence, soluble in hot HCl and gives acicular crystals of $PbCl_2$ upon cooling. The adamantine luster, usual twinning, and effervescence in dilute HNO_3 differentiate it from anglesite.
Occurrence: The most common secondary lead mineral in the oxidized portions of lead deposits; associated with the secondary minerals of the common metals. Notable crystals from Organ Mountains, New Mexico; Sardinia; Tsumeb, South-West Africa; Broken Hill, New South Wales, Australia; Coeur d'Alene district, Idaho; and also from Wheatley Mines, Phoenixville, Pennsylvania; Joplin, Missouri; Leadville, Colorado; Flux Mine, Pima County, and Red Cloud Mine, Yuma County, Arizona.
Uses: A common ore mineral of lead. Reticulated specimens are sought after by collectors for cabinet specimens.

Galena *(illus. pp. 33, 36, 130, 137)*

Isometric, hexoctahedral; usually in cubes; occasionally in octahedrons, often modified; also as granular compact masses. Overgrowths of octahedrons in parallel position on cubes are common. Sulfide of lead, PbS, containing 86.6 percent lead. Silver is frequently in galena. Minute amounts of argentite, tetrahedrite, arsenic, and antimony may be present with the galena. Bluish gray to dark metallic gray; L metallic; S medium gray; H 2.5; SG 7.58; C cubic, perfect, easy; Fr subconchoidal; very brittle; opaque.
Tests: BB F 2; on charcoal yields a metallic button in RF, a yellow sublimate, PbO, near the assay; on plaster tablet with iodine flux yields a lemon yellow sublimate. Decomposed by HNO_3 yielding $PbSO_4$; with HCl gives separation of $PbCl_2$ in minute, white crystals, which are soluble in the hot solution. The high SG, perfect and easy cubic cleavage, low hardness, and blowpipe reactions identify it.
Occurrence: One of the most common minerals in fissure veins or other kinds of mineral deposits. There are thousands of occurrences that have produced fine crystals; probably the most prolific is the area around Joplin, Missouri, where galena is associated with excellent crystals of sphalerite, calcite, dolomite, and marcasite. Other important lead-producing areas are southeastern Missouri; Idaho; Utah; Arizona; Leadville, Colorado; Mineral Point, Wisconsin; Galena, Illinois; Rossie and Ellenville, New York; Kimberley, British Columbia; Bolivia; Peru; England; Germany; Bohemia; and Broken Hill, New South Wales, Australia. The most important ore mineral of lead.

Clay minerals (top)—feldspar (Maine), kaolinite (Cornwall, England), kaolinite (Georgia)—and objects made with clay: Chinese incense burner, English porcelain egg coddler, English bone-china tea cup.

Mimetite

Hexagonal, bipyramidal; usually rounded with curved faces, sometimes as skeleton crystals; also in crusts. A chlorarsenate of lead, $Pb_5(AsO_4)_3Cl$, containing 69.5 percent lead. Mimetite is isomorphous with pyromorphite; hence the mineral may contain appreciable amounts of phosphorus. White to brown to red; L subadamantine; S white; H 3.5 to 4; SG 7 to 7.3; C {10$\bar{1}$1} in traces; Fr subconchoidal; brittle.

Tests: On charcoal fuses easily, yields a metallic button in OF; gives arsenic fumes with a garliclike odor; soluble in HNO_3, yields tests for chlorine and lead; in CT with powdered charcoal gives an arsenic mirror. Its habit places it in the pyromorphite series; it can usually be distinguished by test for arsenic. The presence of phosphorus (pyromorphite) or vanadium (vanadinite) should be checked; if these are absent the mineral is mimetite.

Occurrence: A secondary mineral after lead and arsenic minerals in the oxidized parts of mineral deposits. Some localities are in Saxony, Bohemia, England, Scotland, Australia, Pennsylvania (especially Wheatley Mine, Phoenixville), Arizona, California, Utah (especially Eureka), and Nevada. An ore mineral of lead but rarely of importance.

Linarite

Monoclinic, prismatic; slender crystals, or as groups, crusts, or aggregates; commonly twinned on {100}. A basic sulfate of lead and copper, $PbCuSO_4(OH)_2$, with 15.9 percent copper, and 51.5 percent lead. Deep azure blue; L subadamantine; S pale blue; H 2.5; SG 5.3 to 5.4; C {100} perfect; Fr conchoidal; brittle; transparent on thin edges.

Tests: On charcoal in RF yields a metallic button; CT decrepitates, turns dark, and yields some water. HNO_3 decomposes the powder with separation of lead sulfate and formation of a green solution, which turns blue (copper) upon adding ammonia. Association with lead minerals, lack of effervescence in acids, and incomplete solubility in nitric acid are diagnostic characters. A similarly appearing mineral, diaboleite, is completely soluble in nitric acid.

Occurrence: Widespread in oxidized copper deposits as a secondary mineral. Found in the copper-lead mining areas of Europe, South-West Africa, Chile, Peru, Argentina, Montana, Idaho, and British Columbia. Also in the Tintic district, Utah; Cerro Gordo district, Inyo County, California; Mammoth Mine, Tiger, Arizona; and Blanchard Mine, Socorro County, New Mexico. An accessory ore mineral of copper and lead.

Pyromorphite *(illus. p. 136)*

Hexagonal, bipyramidal; prismatic crystals often with basal terminations in nearly parallel groups of prisms, tapering to a point; individual crystals often skeletonized and showing zoning. A chlorophosphate of lead, $Pb_5(PO_4)_3Cl$, containing 75.4 percent lead. Isomorphous with mimetite and vanadinite. Light to dark green to brownish red; L subadamantine; S white; H 3.5 to 4; SG 7 to 7.1; C pyramidal in traces; Fr subconchoidal; brittle.

Tests: BB on charcoal fuses easily at 1.5 to a globule, crystallizes into a polyhedral form upon cooling. Fused with Na_2CO_3 on charcoal yields a metallic button of lead; CT a white sublimate of lead chloride; soluble in HNO_3 and KOH; a few drops of the dilute HNO_3 solution when added to ammonium molybdate gives a yellow precipitate (phosphorus). When added to a sodium phosphate bead saturated with CuO it yields an azure blue flame (chlorine). Pyromorphite is similar only to the other members of the isomorphous group; the color and blowpipe tests will usually separate them.

Occurrence: A secondary mineral in association with other metallic elements of the primary minerals. Among the best-known localities are Ems, Germany; Broken Hill, New South Wales, Australia; and Phoenixville, Pennsylvania. Numerous other localities are in Europe; the Ural Mountains; England; British Columbia; Davidson County, North Carolina; Coeur d'Alene district, Idaho; Galena and Carbonate, South Dakota; Colorado; and Hampshire County, Massachusetts.

Uses: An accessory ore mineral of lead. Crystal groups are prized as cabinet specimens because of their unique clusters and beauty.

LITHIUM (Li)

Lithium is a white metal, first discovered in 1817. It is extremely light, with a SG of 0.53, or about half that of water. Lithium could float on water, but it is so reactive that it would tend to decompose the water explosively. It is therefore usually stored in kerosene. Lithium is used in steelmaking, special greases for heavy duty bearings, ceramic fluxing, enamels, and in the manufacture of aluminum. Lithium is also used in batteries. There are about 140 lithium-bearing minerals, of which only a few are important.

Spodumene

Monoclinic, prismatic; crystals in stout prisms sometimes of enormous size; vertically striated; also massive. $LiAl(SiO_3)_2$, containing 3.9 percent lithium. Grayish white, yellowish, green, and lilac; L vitreous; H 6.5 to 7; SG 3.13 to 3.20; C {110} perfect; Fr uneven splintery to subconchoidal; brittle; transparent to translucent.

Tests: BB becomes white and opaque, intumesces, imparts a red color to the flame (lithium); fuses at 3.5 to a clear glass. Distinguished by perfect {110}

cleavage, medium-high SG, and red flame BB.
Occurrence: A common mineral in the sodium-rich pegmatites (many crystals more than twenty feet long have been mined); occurs in Sweden; Africa; Brazil; Pala district, California; Etta Mine, Pennington, South Dakota; near Stony Point, North Carolina; and Sterling, Goshen, Chesterfield, and Huntington, Massachusetts.
Uses: The chief ore mineral of lithium. The transparent, colored varieties are prized as gems.

Amblygonite

Triclinic; in large indistinct crystals, compact, massive, also globular, reniform. A fluo-phosphate of aluminum and lithium, $LiAl(F,OH)PO_4$, lithium oxide about 10 percent. Grayish white, yellowish, brownish, various green shades; H 3; SG 4.1 to 4.4; L adamantine to vitreous; Fr conchoidal to even; brittle.
Tests: CT gives water; F 2 with intumescence; colors flame red (lithium); moistened with H_2SO_4 gives a bluish green color to flame. Fine powder dissolves easily in H_2SO_4 and slowly in HCl. Differs from the feldspars, carbonates, and sulfates by easy fusibility and red flame; from spodumene by acid water in CT.
Occurrence: Occurs in the granite pegmatites in association with other lithium minerals. Found in Australia, France, South Africa, Connecticut, California, and Oxford County, Maine. Also in South Dakota at the Hugo, Ingersoll, and Peerless mines, Keystone, and Beecher Lode, Custer County. An ore mineral of lithium.

Lepidolite

Monoclinic; in aggregates of short prisms, also in cleavable plates, masses of flat crystal plates, and coarse to fine granular micaceous masses. $(OH,F)_2KLiAl_2Si_3O_{10}$; the composition is variable from one locality to another. Rose red to violet gray or weak to dense lilac, yellowish, grayish white; H 2.5 to 4; SG 2.8 to 3.3; L pearly; C {001} eminent.
Tests: CT yields acid water; F 2 to 2.5; BB with intumescence; colors flame red (lithium); partially decomposed by acids; after fusion yields a gelatinous precipitate with HCl. Distinguished by eminent cleavage, purplish color, and blowpipe reactions.
Occurrence: A common mineral of the sodium-lithium phases of the granite pegmatites in Europe; South Africa; Pala district, San Diego County, California; Ingersoll Mine, Keystone, South Dakota; and Hebron, Buckfield, and Paris, Maine. An ore mineral of lithium.

MAGNESIUM (Mg)

Magnesium, a white metal, is very abundant in the earth's crust. It is very light in weight (SG 1.74), and is used extensively in lightweight alloys. Magnesium resists corrosion and can be easily formed into structural shapes. The powder burns readily with an intense white light, useful in flares and fireworks. Many household items and machine parts are die-cast magnesium alloys. More than 150 minerals contain magnesium, but few are economically important as sources of the metal.

Magnesite

Hexagonal, rhombohedral; usually in hard, compact, or soft earthy masses; rarely in small, rhombohedral crystals; also in coarsely crystalline masses. Magnesium carbonate, $MgCO_3$, containing 28.8 percent magnesium; some iron may be present as iron carbonate. White, grayish white to brown; L vitreous; H 3.5 to 4.5; SG 3 to 3.12; C {1011} perfect; Fr conchoidal; brittle; transparent to opaque.
Tests: Slightly acted upon by cold acids; in powder dissolves in warm HCl with effervescence; gives tests for magnesium in absence of much calcium; BB infusible, glows. Distinguished by test for magnesium. Hard, compact varieties resemble chert but are decidedly softer. Difficult to distinguish from dolomite in the field.
Occurrence: An alteration product of magnesium-rich rocks. Large deposits on the island of Euboea, Greece, and along the western slope of the Sierra Nevada and coast ranges of California; Chewelah, Washington; the Muddy Valley district and Luning, Nevada; Bahia, Brazil; Austria; India; Manchuria; Korea; and South Africa.
Uses: Most important use is for refractory magnesite bricks for lining metallurgical furnaces and in the manufacture of various chemical compounds.

Dolomite *(illus. p. 17)*

Hexagonal, rhombohedral, in rhombohedral crystals, which are frequently curved, commonly quite small, and sometimes grouped into saddle-shaped forms. Occurs worldwide in the form of dolomitic limestone, fine- to coarse-grained. $CaMg(CO_3)_2$, intermediate between magnesite and calcite, containing theoretically 13 percent magnesium. Usually white or gray, other colors depend upon impurities; L vitreous to pearly; H 3.4 to 4; SG 2.85; C rhombohedral; perfect; brittle.
Tests: BB infusible; not reacted upon by cold acid; the powder readily dissolves with effervescence. Resembles calcite but is harder and heavier. Calcite dissolves readily in cold, dilute acids.
Occurrence: Found in Keokuk, Iowa; Lockport and Niagara Falls, New York; Hoboken, New Jersey; Hiddenite, North Carolina; and Joplin district, Missouri.

Periclase

Isometric in cubes, octahedrons, and grains with cubic cleavage;

MgO; H 6; SG 3.67 to 3.90. Occurs in limestone at Mount Somma, Vesuvius, Italy; Varmland, Sweden; Organ Mountain, New Mexico; and Wet Weather and Jensen quarries, Riverside County, California. Too rare to be of commercial use.
Tests: Same as tests for magnesium.

Spinel

Isometric, usually as octahedrons, truncated by dodecahedrons; twins {111} common, from which is derived the law of spinel twins. A magnesium aluminate, $MgAl_2O_4$; iron and manganese may be present. Yellow, brown, black, colorless, green, blue, and red; C {111} imperfect; Fr conchoidal; brittle; H 8; SG 3.5 to 4; L vitreous; S white.
Tests: BB infusible; slowly soluble in the fluxes; salt of phosphorus yields a bead white when hot and a faint chrome green when cold. Decomposed by fusion with potassium bisulfate. Distinguished from magnetite, which is magnetic. Zircon has higher SG; corundum is harder; garnet is softer and easily fusible.
Occurrence: An accessory in many basic rocks, both igneous and metamorphic. A residual mineral in gravels in Thailand and Sri Lanka as gemstones. Excellent crystals have been found at many places: in a band of granular limestone from Amity, New York, to Andover, New Jersey; also in Culsaggee Mines, near Franklin, North Carolina. Not abundant enough for commercial use except for the colored varieties, which make excellent gemstones.

MANGANESE (Mn) *(illus. p. 140)*

Manganese is one of our most useful metals. It was employed 5,000 years ago by Egyptian glassmakers to remove undesirable coloring matter. The metal was finally isolated in 1774, by charcoal reduction of manganese dioxide.

Metallic manganese is gray and resembles cast iron. It is very good for removing sulfur and oxygen from steel, and about ten pounds of manganese are required to make a ton of steel. Manganese is also used as an alloying element in making armor plate and armor-piercing projectiles, structural steel, car wheels, burglar-proof safes, railroad rails, ore-crushing machinery, teeth for earth-moving equipment, and many other applications. Manganese is also consumed in making dry-cell batteries. Manganese compounds have myriad other industrial uses.

There are more than 150 manganese-bearing minerals, but only a few serve as ores. Manganese as an impurity produces pink colors in various other minerals, and two manganese minerals, rhodonite and rhodochrosite, are used widely for gems and decorative stone applications. Most other manganese minerals tend to be black.

Franklinite

Isometric, hexoctahedral; crystals common and are principally octahedrons, sometimes cubes, often modified by other forms. An oxide of iron, zinc, and manganese $(Fe,Zn,Mn)(Fe,Mn)_2O_4$. Black; L brilliant metallic; S reddish brown; H 5.5 to 6.5; SG 5.07 to 5.22; C none; Fr uneven; brittle. If high in iron, the mineral is magnetic to a variable degree.
Tests: BB infusible; yields tests for manganese, iron; also for zinc with cobalt nitrate. Distinguished as a member of the spinel group; its crystal forms and association with calcite, zincite, and willemite are diagnostic features; resembles magnetite, which is strongly magnetic; chromite is associated with basic igneous rocks.
Occurrence: In a strongly metamorphosed sedimentary rock originally consisting of concentrated weathered products of minerals exceedingly rich in calcium, iron, manganese, and zinc. The mineral is rare beyond the deposits at Franklin and Sterling Hill, New Jersey. These areas are famous for their rich mineral content of more than a hundred species.
Uses: An important ore mineral of zinc and manganese. Many fine specimens greatly prized by collectors have come from the deposits at Franklin and Sterling Hill, New Jersey.

Manganite

Monoclinic, prismatic; crystals short to long prismatic, vertically striated; also columnar. Twinning {011} and {100}. A basic oxide of manganese, MnO (OH), containing 62.4 percent manganese. Gray to black; L submetallic; S reddish brown to black; H 4; SG 4.33; C {010} perfect; Fr uneven; brittle.
Tests: BB infusible; with small amount of the mineral, it reacts for manganese with the fluxes; dissolves in HCl with evolution of chlorine; yields much water in CT. Crystals are distinctive; difficult to distinguish from other manganese oxides; streaks, SG, and H are different.
Occurrence: A low-temperature vein mineral. Fine specimen groups from Ilefeld, Harz Mountains, Germany; also in France, England, and Scotland. In the United States at Sterling Hill, New Jersey; Cartersville, Georgia; Jackson, Maine; Negaunee, Michigan; and elsewhere in the Lake Superior iron district, in both crystal groups and massive.
Uses: An ore mineral of manganese and an attractive cabinet specimen.

Psilomelane

Orthorhombic; in concretionary forms, also as powdery and earthy material. A basic oxide of barium and manganese, $BaMn(MnO_2)_8(OH)_4$; small amounts of impurities are usually present. Black; L submetallic; S brownish black; H 5 to 6; SG

3.7 to 4.7; Fr subconchoidal; brittle; opaque.
Tests: BB on charcoal infusible; CT yields water; soluble in HCl with evolution of chlorine; gives manganese reaction with the fluxes; the HCl solution yields a precipitate of $BaSO_4$ upon addition of a soluble sulfate. Separated from other manganese minerals by precipitation of $BaSO_4$. The other physical and blowpipe reactions are important identifying properties.
Occurrence: A mineral of secondary origin derived from the weathering of manganese minerals and associated with primary and other secondary manganese minerals, also secondary iron minerals. Found in Lake Superior hematite deposits, Negaunee, Michigan, and elsewhere.
Use: An ore mineral of manganese and probably second in importance only to pyrolusite.

Pyrolusite

Tetragonal, ditrigonal, bipyramidal; prismatic and equant crystals; usually massive, sometimes fibrous in concentric growths producing reniform developments, also granular and powdery growths. Many fractures in rocks contain dendritic growths of pyrolusite; where similar growths occur in banded chalcedony the material is known as moss agate. Manganese dioxide, MnO_2, containing 63.2 percent manganese. The mineral is often an alteration product of other manganese minerals; dark gray to black; L metallic; S black; H 6 to 6.5; SG 4.4 to 5.2; C {110} perfect; Fr uneven; brittle; opaque.
Tests: BB infusible; CT may give a small amount of water; also yields oxygen; soluble in HCl with evolution of chlorine; a minute quantity imparts a purplish color to the borax bead in the OF; bead becomes colorless in RF. Distinguished by blowpipe tests. To distinguish the granular and powdery varieties of manganese minerals, the X-ray diffraction patterns are needed.
Occurrence: Forms under oxidizing conditions and is present in many countries. In the United States good specimens originate in the Cayuna Range in Minnesota; in the Lake Superior area; in Batesville, Arkansas; and in Vermont.
Uses: Pyrolusite is the most important ore mineral of manganese. The mineral is used as a deoxidizer in many chemical processes; to decolorize glass, as a dryer in the manufacture of paints and varnishes, in dry batteries; for the making of chlorine, bromine, oxygen, and numerous other applications.

Rhodochrosite *(illus. p. 140)*

Hexagonal; hexagonal-scalenohedral; crystals are usually simple rhombohedrons, rarely scalenohedrons; also globular, botryoidal, compact, massive, coarsely granular, sometimes in concentric bands; distinct rhombohedral cleavage. Manganese carbonate, $MnCO_3$; magnesium, iron, and calcium are often present in varying amounts. From pink to rose red, sometimes yellowish to brownish; L vitreous; S white; H 3.5 to 4; SG 3.5 to 3.7 but varies with the amount of other elements present; C {1011} perfect; Fr conchoidal; brittle; transparent.
Tests: BB infusible, becomes black, decrepitates strongly; soluble in warm HCl with effervescence; gives the reaction for manganese with the fluxes. Distinguished by rose red color, rhombohedral cleavage, and greater softness as compared with rhodonite.
Occurrence: Associated with lead, copper, zinc, and silver minerals in low-temperature veins; also in contact metamorphic deposits. From many localities in Europe, USSR, and South America. Mineralogically, the finest specimens come from Colorado (in particular, Park County) as large, pink rhombohedrons. Small but excellent scalenohedrons of a deep red color from Westphalia. Other United States localities are Butte, Montana; Austin, Nevada; Negaunee, Michigan; Franklin, New Jersey; and Branchville, Connecticut.
Uses: An ore mineral of manganese. The single crystals and groups of fine rose red crystals are highly prized by collectors.

Rhodonite

Triclinic, pinacoidal; crystals tabular and short, prismatic habit, also in very sharp crystals of minute size; usually fine-grained and massive. A manganese metasilicate, $MnSiO_3$, containing 41.8 percent manganese; iron, magnesium, calcium, and rarely zinc are present. The calcium-rich variety is called bustamite and the zinc-rich variety is known as fowlerite. Flesh red to bright pinkish red, varies to grayish when impure, turns black by weathering. L vitreous; S white; H 5.5 to 6.5; SG 3.4 to 3.7; C {110} perfect; Fr uneven; tough; transparent to translucent.
Tests: BB blackens and fuses easily to a black glass with slight intumescence; gives reaction for manganese with the fluxes. Characterized by crystal forms. Tephroite, Mn_2SiO_4, is darker red, otherwise difficult to distinguish except by X-ray diffraction. Rhodochrosite is much softer and soluble in warm, dilute acids.
Occurrence: An occasional mineral in the silver-lead veins as at Broken Hill, New South Wales, Australia. The outstanding occurrence is at Franklin, New Jersey. Massive rhodonite occurs in the Ural Mountains. Excellent material, although in small amounts, has come from Cummington and Plainfield, Massachusetts. Other localities are Washington and Sweden.
Uses: An ore mineral of manganese, but rarely used as such. The colorful material is used for semiprecious stones and decorative purposes.

MERCURY (Hg) *(illus. p. 141)*

Mercury was known and used by the ancient Chinese and Hindus as early as 1600 B.C. Flasks of mercury were buried in Egyptian tombs. The Greeks imported mercury from Spain as early as 700 B.C. Mercury is the only metal that is liquid at temperatures normal for the temperate zone, MP −38.87° C. Liquid mercury, SG 13.55, is also one of the heaviest metals. Applications include thermometers, barometers, and scientific instruments, vapor lamps, explosives, poisons, and insecticides and electrical switches. Amalgams, which are mercury alloys, are used in recovering gold and silver and in dental work. Only one mercury mineral—cinnabar—is of economic importance, although a few rare species—livingstonite, eglestonite, terlinguaite, and others—

Top: Meerschaum—or sepiolite, a claylike mineral—from Turkey, and a calabash pipe with meerschaum bowl. *Above:* Richterite, a mineral in the amphibole group, from Quebec. *Opposite:* Diopside, a pyroxene, also from Quebec. Both of these are important rock-forming minerals.

are of some interest to collectors. Native mercury occurs in cinnabar deposits.

Cinnabar *(illus. p. 141)*

Hexagonal, trigonal, trapezohedral; crystals uncommon; crusts common; usually granular, massive, or earthy, sometimes as fine needles. A sulfide of mercury, HgS, containing 86.2 percent mercury. Scarlet, bright red, vermilion, darker with impurities; transparent crystals are rich red; L adamantine; S scarlet; H 2 to 2.5; SG 8.1; C prismatic, perfect; Fr conchoidal.
Tests: BB pure material is entirely volatile; CT with Na_2CO_3 gives a sublimate of metallic mercury; OT gives sulfurous fumes and metallic mercury sublimate in minute globules on cold glass walls. Distinguished by color, streak, high SG, and blowpipe reactions. Cuprite sometimes resembles cinnabar but its crystals are isometric and its streak is brownish red. BB cuprite yields a malleable button of copper.
Occurrence: Usually in areas of recent igneous activity, where it is deposited in veins and as disseminations in porous rocks and in hot-spring deposits. Turkestan and China have produced mercury for centuries; also the Balkan countries westward into Spain, especially near Belgrade and Idria, Yugoslavia; Italy; and Almaden, Spain. In the United States mercury is produced at New Almaden and New Idria, California; Terlingua, Texas; Pike County, Arkansas; Idaho; Nevada; and Oregon. The mines at Almaden, Spain, have been worked for over 2,500 years and are still highly productive.
Uses: Cinnabar is the only important ore mineral of mercury. From the standpoint of the mineral collector the groups of twinned cinnabar crystals from Hunan Province, China, are among the most spectacular specimens known. Important deposits elsewhere likewise furnish many specimens.

MOLYBDENUM (Mo)

In 1778, Karl Scheele, a Swedish chemist, discovered that nitric acid reacted with a mineral strongly resembling graphite to yield a white powder he called "molybdic acid." This "acid" was reduced by Hjelm four years later, yielding a new metal, molybdenum. Molybdenum is a white metal that is very hard. It melts at 4748° F., a melting point exceeded only by tantalum, tungsten, osmium, and rhenium. Molybdenum carbide is remarkably hard, and molybdenum in steel adds hardness, toughness, strength, and elasticity. Molybdenum is also used in electrical devices, incandescent lamps, pottery, fabrics, analytical chemistry, and various alloys. Only a few minerals are sought by collectors; the chief ore is molybdenite.

Molybdenite

Hexagonal, dihexagonal bipyramidal; occurring chiefly as tabular hexagonal plates or in short prisms or pyramids, usually occurring as foliated or scaly masses in narrow seams. A sulfide of molybdenum, MoS_2, usually free from other elements. Medium metallic gray; L metallic; S bluish metallic gray on paper; H 1 to 1.5; SG 4.6 to 4.7; C {0001} perfect; laminae very flexible but not elastic; opaque; unctuous to the touch.
Tests: BB infusible; on charcoal gives a white sublimate at a distance with yellow to red near the assay. The white coating when touched with the reducing flame becomes azure blue; with sodium phosphate becomes green in the RF; decomposed by HNO_3, leaving a residue of MoO_3, which is soluble in NH_4OH; the addition of a small amount of Na_2HPO_4 gives a yellow precipitate. Graphite is the only similar mineral but is much lighter. Blowpipe tests separate the minerals.
Occurrence: Molybdenite is an accessory mineral in acid igneous rocks such as granites; also in acid pegmatites and high-temperature veins associated with topaz, wolframite,

scheelite, fluorite, and cassiterite. It is widespread but rarely segregated in sufficient amounts to be of economic importance. In the United States chiefly from the Rocky Mountain states; also from Chile, Norway, and Canada. The best specimens probably come from Pontiac County, Quebec. Other localities are Climax Mine, Lake County, Colorado; Blue Hill Bay, Maine; Westmoreland, New Hampshire; and Haddam, Connecticut.
Uses: The chief ore mineral of molybdenum and its compounds. It is sometimes used in special lubricants.

Wulfenite *(illus. pp. 17, 96, 143)*

Tetragonal, pyramidal, usually in tabular crystals, often very thin with large basal planes; occasionally with a pyramidal and prismatic habit. The presence of the third order pyramid faces sometimes gives the crystals a twisted appearance; also massive, coarse- and fine-grained. A lead molybdate, $PbMoO_4$; tungsten and calcium in small amounts may proxy for molybdenum and lead; chromium may be present in the red varieties. Yellow, orange, to bright red and brown; L adamantine; S white; H 2.5 to 3; SG 6.5 to 7.5; C pyramidal, good; Fr subconchoidal; very brittle; transparent to translucent.
Tests: BB decrepitates; OF with borax colorless; RF opaque; OF with salt of phosphorus yellowish green when hot, almost colorless when cold; with Na_2CO_3 on charcoal gives button of lead; heated in HCl with a piece of zinc gives the solution a deep blue color. The crystal forms and color usually identify the mineral; the red variety may resemble crocoite, but the crystal forms and blowpipe tests distinguish the two species.
Occurrence: A secondary mineral in the oxidized parts of mineral deposits containing lead and molybdenum. Found in Germany, Bohemia, Sardinia, and North Africa; localities in Australia and Mexico. In the United States from numerous localities in the Southwest, especially Red Cloud Mine, Yuma County, Arizona; Hillsboro, New Mexico; Lucin district, Utah; Loudville, Massachusetts; and Phoenixville, Pennsylvania.
Uses: A minor ore of lead and molybdenum. Mineralogically, wulfenite ranks as one of the most desirable of all minerals because of the very showy character of its specimens. By far the finest of these are the gorgeous red crystal groups from the Red Cloud Mine, Yuma County, Arizona.

Powellite

Tetragonal, bipyramidal; usually as bipyramids and thin basal plates, also massive, powdery, and as crusts. Calcium molybdate with calcium tungstate; $Ca(Mo,W)O_4$. Straw yellow, greenish yellow, brown to nearly black; L subadamantine; H 3.5 to 4; SG 4.2 to 4.6; Fr uneven; transparent; fluoresces in shades of yellow in ultraviolet light.
Tests: F 5 forming a gray mass; decomposed by HCl and HNO_3. A drop of the chloride solution evaporated on a streak plate leaves a blue residue. Gives a test for molybdenum with salt of phosphorus bead. The HCl solution made alkaline with ammonia gives a precipitate with ammonium oxalate (calcium). Distinguished from scheelite by its yellow fluorescence and lower SG; blowpipe tests show the presence of molybdenum.
Occurrence: Powellite is a secondary mineral from the alteration of molybdenite. Excellent crystals have been found in the South Hecla and Isle Royale Mines, Houghton County, Michigan. Also in Tonopah, Nevada; Llano County, Texas; New Mexico; Arizona; and Kern County in California. The mineral is found also in the Ural Mountains, the Balkans, Turkey, and North Africa. A minor ore of molybdenum and tungsten.

NICKEL (Ni)

Centuries ago miners in Saxony

(Germany) were troubled by a reddish mineral in their ores that looked like copper, but would not yield metallic copper. They called the mineral "cupfer nickel," meaning "copper demon." In 1751, the Swedish chemist Cronstedt succeeded in extracting from this mineral a white metal that he named "nickel." Today, nickel is a well-known and widely used metal. The United States five-cent coin is made of nickel-copper alloy. Pure nickel is strongly magnetic. Nickel is used primarily in steelmaking and for alloys, such as stainless steel. Monel metal is a nickel-copper alloy with steel-like strength and toughness, immunity to rust, and resistance to corrosive chemicals, including salt, alkalis, and acids. Nickel, because of its many uses, is therefore a strategic metal, vital to all of industry.

More than forty minerals contain nickel as an essential element. Only a few of these are important ore minerals. Millerite is popular among mineral collectors because of its spectacular, yellow, needlelike crystals.

Niccolite

Hexagonal; usually massive. A nickel arsenide, NiAs, with 43.9 percent nickel; may contain some iron, cobalt, and antimony. Pale copper red; L metallic; S pale copper red; H 5.5; SG 7.3 to 7.7; Fr uneven; brittle.
Tests: Yields a sublimate of arsenic trioxide in OT. Soluble in aqua regia. Commonly associated with other nickel, cobalt, and silver minerals.
Occurrence: Austria, Bohemia, Saxony, France, and Colorado (especially Copper King Mine, Boulder County); Stanislaus Mine, Calaveras County, California; Franklin, New Jersey; with silver and cobalt ores at Cobalt and Thunder Bay, Ontario. An ore mineral of nickel.

Millerite *(illus. p. 143)*

Hexagonal, rhombohedral; usually in radiating needlelike crystals; globular; elastic. A nickel sulfide, NiS, containing 64.7 percent nickel. Color brass yellow; S greenish black; H 3 to 3.5; SG 5.3 to 5.65; cleavage parallel to first order pyramid; brittle.
Tests: BB on charcoal fuses to a globule; with fluxes gives a violet bead in OF, becoming gray in RF from reduced nickel; on charcoal gives a solid metallic and magnetic mass.
Occurrence: In cavities and among the crystals of other minerals in some geodes; with nickel, cobalt, silver ores of Bohemia; Glamorgan, Wales; Antwerp, New York; Gap Mine, Lancaster County, Pennsylvania; geodes at Keokuk, Iowa; and from the Sudbury District, Ontario.

PLATINUM GROUP: PLATINUM (Pt), PALLADIUM (Pd)

Platinum and palladium are white, silvery metals with extremely high melting points and high specific gravities. Both are ductile and malleable, absorb gases, and are chemically inert. Related metals considered part of the platinum group include osmium (Os), iridium (Ir), ruthenium (Ru), and rhodium (Rh). A chief use of these metals is in jewelry and petroleum refining, and platinum has at times also been used in coinage. The platinum metals are rare, occur in native form as nuggets and grains in placer deposits, and are also associated with chromium deposits.

The platinum metals occur in placer deposits in Colombia, Alaska, and the Ural Mountains, which were the sole early sources. At present, platinum metals are obtained in quantity from the copper-nickel ore deposits in basic rocks at Sudbury, Ontario, and from the ultra basic rocks of the Transvaal, South Africa. Also found at Rambler Mine, Encampment, Wyoming, and in Franklin, North Carolina.

Platinum

Isometric. Metallic grayish white; H 4 to 4.5; SG 21.45; ductile and malleable. Melts at 173.5° C; soluble only in aqua regia; inert in the atmosphere; absorbs gases. Found at Cape Blanco, Oregon; Platinum,

Opposite: A pewter pitcher, made from an alloy of antimony (bottom l.) and tin (bottom r.). Top, the mineral sources of these metals—stibnite (l.), from Japan, and cassiterite, from Bolivia. *Above:* Bars of pure titanium, with its ores, rutile (top l.) and ilmenite (top r.).

Brilliant asterism—a starlike effect produced by inclusions—on a piece of phlogopite, a rock-forming mineral in the mica group.

Alaska; and Burke and Rutherford counties, North Carolina.

Palladium

Isometric. Metallic grayish white; H 4.5 to 5; SG 12; ductile and malleable. Melts at 1553° C; soluble in hot HNO_3 and aqua regia; inert in the atmosphere; absorbs hydrogen. Found in Colombia and Minas Gerais, Brazil.

Sperrylite

A diarsenide of platinum with 56 percent platinum. L metallic white; S black; H 6 to 7; SG 10.6; Fr conchoidal; brittle. Recognized by its great weight, crystallization, and brilliant luster. It is the principal source of platinum in the copper-nickel ores of Sudbury, Ontario, and the basic rocks of South Africa.

SELENIUM (Se)

Selenium is an acid-forming element with properties between those of sulfur and tellurium. Although it occurs in twenty or more minerals, none is known to occur in deposits by itself. Most of its minerals are selenides of bismuth, silver, copper, mercury, or sulfoselenides of one or more of these metals, and occur in association with sulfides, arsenides, and antimonides of the nonferrous metals; also in association with pyrite.

The addition of selenium to steel produces an alloy with increased machinability. It is used in the rubber industry and in glassmaking. Since its electrical conductivity is low in the light and high in the dark, it is useful in photographic exposure meters.

SILVER (Ag) *(illus. pp. 34, 143, 188)*

The use of silver extends far back into prehistoric times. It was known by all the great nations of antiquity, and served as a medium of exchange even in the days of Abraham. Recovery processes were known as far back as 4,000 years ago. Ancient silver came from Laurium, Greece; Joachimsthal, Bohemia; Freiberg, Saxony; and later Peru and Mexico. The Bohemian mines were so rich that a mint was established to strike a coin called the *Joachimsthaler,* later contracted to *thaler,* which still later provided our word "dollar."

Silver was not used in industrial applications until the past few decades. It is an excellent electrical conductor and is used widely in electrical devices. Photographic chemicals are the largest single industrial use. Some silver is used in alloys and solders, electroplating, dental work, bearings, catalysts, and batteries. Of course, a major application is jewelry, flatware, holloware, coins, medallions, and high-reflectivity mirrors.

Silver is an important element in about sixty minerals, of which several are popular among collectors.

Native Silver

Isometric; usually in acicular forms; arborescent and coarse- to fine-wire forms; also massive. A metallic white color; L metallic; H 2.5 to 3; SG 10.1 to 11.1; C none; ductile; malleable. Native silver may contain up to 10 percent gold, a little copper, sometimes antimony, bismuth, and mercury.

Tests: BB fuses easily to a metallic white globule; in OF gives a faint red coating of silver oxide; soluble in HNO_3 and precipitated by a piece of copper. A chloride precipitates a white, curdy silver chloride. Distinguished by its color and SG.

Occurrence: Usually a secondary mineral found in the upper parts of deposits containing silver minerals. Primary silver occurred in large amounts associated with basic rocks at Cobalt, Ontario, where cobalt and nickel minerals were common and abundant associates. Outstanding native silver specimens have been found at Kongsberg, Norway. It occurs in several places in Mexico, in the western United States, and with copper on the Keweenaw Peninsula, Michigan. Unusually large masses of silver were found as float and in the mines at Cobalt, Ontario. An ore of silver.

Argentite

Isometric, octahedrons and cubes; sometimes arborescent, also massive. A silver sulfide, Ag_2S, containing 87.1 percent silver. Blackish lead gray; H 2 to 2.5; SG 7.2 to 7.4; perfectly sectile.

Tests: BB on charcoal fusing with intumescence in OF; yielding a globule of silver. OT gives off sulfurous fumes. Distinguished from other black minerals by being readily cut with a knife, also by giving a silver button on charcoal.

Occurrence: Argentite occurs in low-temperature mineral deposits associated with copper, bismuth, zinc, lead, nickel, cobalt, and other silver minerals. Occurs in the mines of the silver districts of eastern Europe, the Andes Mountains of South America, Mexico, Australia, western United States (particularly in Butte, Montana; Aspen and Leadville, Colorado; Comstock lode and Tonopah, Nevada), and Ontario, Canada. An important ore mineral of silver.

Pyrargyrite

Ruby silver ore; rhombohedral, hemimorphic; crystals usually prismatic, also massive, compact. A double salt of silver and antimony, $3Ag_2S \cdot Sb_2S_3$, containing 59.9 percent silver; sometimes small amounts of arsenic may be present. Splinters are deep red by transmitted light; L metallic to adamantine; S purplish red; H 2.5; SG 5.8 to 5.9; Fr conchoidal; brittle.

Tests: CT fuses and gives a reddish sublimate of antimony oxysulfide; OT sulfurous fumes and white sublimate of antimony oxide; BB on charcoal, with fluxes yields a globule of silver in RF; decomposed by HNO_3.

Occurrence: A primary mineral found in the upper parts of mineral deposits; usually asso-

ciated with other silver minerals and with galena, sphalerite, tetrahedrite, chalcopyrite, and quartz. Found in ruby district, Gunnison County, Colorado; Austin and Comstock lode, Nevada; Poorman Mine, Owyhee County, Idaho.

Proustite

Rhombohedral, hemimorphic; in acute scalenohedrons or rhombohedrons, also massive, compact. A double salt of silver and arsenic, $3Ag_2S \cdot As_2S_3$, containing 65.4 percent silver. Scarlet to vermilion; L adamantine; S inclined to aurora red; H 2 to 2.5; SG 5.6 to 5.7.
Tests: CT fuses easily, gives a slight sublimate of arsenic trisulfide; BB on charcoal fuses, gives odor of sulfur and arsenic; with Na_2CO_3 gives a globule of silver in RF. Decomposed by HNO_3 with separation of sulfur.
Occurrence: Under same conditions as pyrargyrite and in many of the richer silver mines. An ore mineral of silver. Found at Red Mountain (San Juan district) and in Georgetown, Colorado; also in Silver City district, Idaho.

STRONTIUM (Sr)

Strontium, a soft white metal, burns with a purplish red flame when heated in air. This gives it use in fireworks and flares where a red flame is required. The SG is about 2.5, the same as aluminum. Strontium occurs in about a dozen minerals, only two of which are important.

Strontianite

Orthorhombic, crystals frequently acicular, pointed, also columnar, fibrous. A carbonate of strontium, $SrCO_3$. White, greenish, yellowish brown; L vitreous; S white; H 3.5 to 4; SG 3.7; C prismatic, nearly perfect; Fr uneven; brittle.
Tests: F only on thin edges, swells up, colors flame red; soluble in HCl; gives a white precipitate with sulfuric acid. Differs from similar white minerals in SG, flame coloration, solubility in acids.
Occurrence: Commonly occurs in veins and cavities in limestone; also as small crystals or grains in metalliferous deposits, normally associated with celestite and calcite. Found in Strontium Hill, California; Woodville, Ohio; Schoharie, New York; Pennsylvania; Texas; South Dakota; and in Carleton County, Ontario. Good crystals come from the Tyrol, Austria; also from Westphalia, Germany, and Strontia, Scotland.

Celestite *(illus. p. 109)*

Orthorhombic; tabular crystals, also granular, massive and fibrous. A strontium sulfate, $SrSO_4$. White, blue, reddish; L vitreous; H 3 to 3.5; SG 3.9; C perfect prismatic and basal; brittle.
Tests: Insoluble in acids. BB may decrepitate; F 3; colors flame red; fused assay on charcoal when treated with HCl and alcohol gives an intensely red flame. Does not effervesce with acids. Distinguished from similar minerals by characteristic crystals, cleavage, high SG, and flame coloration.
Occurrence: In metalliferous veins with galena, sphalerite, and chalcopyrite. Many localities in Europe and England yield the mineral. Found in limestones at Dundas, Ontario; Lockport and Chittenango Falls, New York; West Virginia; Clay Center, Ohio; Kansas; Maybee, Michigan; Lampasas, Texas; Borate, California; and states of the Rocky Mountains area. Used chiefly in chemicals and fireworks.

SULFUR (S) *(illus. pp. 20, 169)*
Sulfur and its chief chemical product, sulfuric acid, are indispensable to modern industry. Sulfuric acid is one of the world's most basic chemical raw materials, employed in making thousands of products. Sulfur is mined in native form from vast deposits, usually associated with salt domes, such as those of the southern United States. Sulfur is also recovered from "sour" natural gas, and occasionally from metallic sulfide minerals, such as pyrite (iron sulfide). Sulfur is a vital material in the manufacture of fertilizers, such as soluble phosphates.

Native sulfur occurs in spectacular crystals in some ore deposits, as well as in volcanic areas, where it has been deposited by sublimation from vapors. Sulfur is a basic constituent of hundreds of minerals.

Sulfur
Orthorhombic crystals, usually as acute pyramids, or thick tabular parallel to the base; also reniform masses, stalactites, and as a powder. Yellow to yellowish brown; L resinous; H 1.5 to 2.5; SG 2.05 to 2.09; C imperfect; Fr conchoidal; transparent to translucent; brittle. Sulfur becomes negative when electrified by friction.
Tests: BB melts at 108° C and burns with a bluish flame to sulfur dioxide with a characteristic irritating odor. It is insoluble in water and acids but dissolves in carbon disulfide. It is distinguished by its color, fusibility, and ability to burn to sulfur dioxide.
Occurrence: Sulfur is derived from many volcanic processes, those on the island of Sicily yielding many noteworthy crystal groups. Other sources are limestones and clay rocks, where it is associated with celestite, calcite, and gypsum. Large deposits are known along the Gulf Coast in Lake Charles, Calsasien Parish, Louisiana, and Freeport, Texas. Also found in Sulfurdale, Utah, and in Cody and Thermopolis, Wyoming.
Uses: Modern industry depends largely on sulfur and its products, especially sulfuric acid, which is essential in many processes. Sulfur is important in the sulfite process of converting wood pulp into paper; in the manufacture of matches, fireworks, insecticides, and gunpowder; in vulcanizing rubber and making pharmaceuticals.

Pyrite *(illus. pp. 17, 64, 75, 90)*
Isometric; crystals usually in cubes, octahedrons, and pyritohedrons; also granular, reniform, globular, stalactitic, and other forms. One of the most common metallic minerals. Pyrite is dimorphous with marcasite, and both may occur in clay and carbonaceous deposits. FeS_2; pale brassy yellow; L glistening metallic, sometimes with mirrorlike surfaces; S greenish black; H 6 to 6.5; SG 4.95 to 5.10; C indistinct; brittle.
Tests: BB fuses at 2.5 to 3; yields sulfur dioxide; becomes magnetic. CT gives much sulfur; insoluble in HCl; powder soluble in nitric acid. Separated from chalcopyrite by its much greater hardness and paler color; from marcasite, which has orthorhombic crystal forms and is whiter in color. Pyrite changes readily into limonite.
Occurrence: Pyrite, the most common sulfide, occurs in all kinds of rocks in sizes from scattered grains to large masses, often in excellent crystals that are sometimes bounded by many crystal forms, which make interesting collector's specimens. Invariably associated with ore deposits. In some localities it carries small amounts of copper and gold and becomes an important ore of these metals. It occurs in fine crystals in many European localities. In the United States good crystals come from Connecticut; New York; the French Creek mines (Chester County) and Cornwall, Pennsylvania; Central City, Colorado; and Bingham Canyon, Park City district, Utah.
Uses: The chief use of pyrite is for its sulfur content; the iron is also recovered. The sulfur of pyrite, along with that of other sulfur-bearing minerals, is recovered from smelters and converted into sulfuric acid.

TIN (Sn) *(illus. p. 161)*
Tin is silvery white and rather soft, malleable, and ductile. Its principal uses are as tinplate on iron, in various alloys, including pewter, in chemicals for dyeing, and—as stannous fluoride—in toothpaste and water treatment. (The prevention of tooth decay, however, is due to the fluorine rather than the tin.)

Tin occurs in about a dozen minerals in small amounts, but only cassiterite is important as ore or to collectors. Bronze, an alloy of copper and tin, antedates recorded history and is found in extremely old burial vaults and archeological sites.

Cassiterite *(illus. pp. 144, 161)*
Tetragonal, prismatic; crystals of prismatic and pyramidal habit are the rule, less frequently in radiating fibrous structure. An oxide of tin, SnO_2, with 78.7 percent tin. Brown or black, sometimes yellow to red; L adamantine to submetallic; S white to brownish; H 6 to 7;

SG 6.8 to 7.1; brittle; transparent to opaque. Distinguished from nearly all nonmetallic minerals by its high SG and its greater hardness.
Occurrence: Occurs in high-temperature granitic veins and pegmatites, in contact metamorphic deposits, and in placer deposits; sparsely at a number of localities throughout the world. Among the richer areas are Cornwall, England; southeastern Europe; Bolivia; and the placer deposits of the Malay Peninsula. Also found at Norway and Hebron, Maine; Irish Creek, Virginia; Harney Peak and Custer City, South Dakota; and Temescal, California.

TITANIUM (Ti) *(illus. p. 161)*

Titanium has been called the modern wonder metal. It is grayish white with a bright luster. Although brittle, titanium adds toughness to its alloys, as well as strength and corrosion resistance. It is used largely in airplanes, jet engines, and industrial applications. Titanium dioxide is a white pigment used extensively in paints, face powders, creams, paper, soap, linoleum, rubber, and drawing and printer's ink. Titanium is found in about fifty minerals as an essential element. Only a few are important enough to be considered ores.

Ilmenite *(illus. pp. 57, 161)*
Hexagonal, trirhombohedral; crystals thick, tabular, and acute rhombohedral; often in thin plates or laminae; also compact massive, granular, and as sand. A titanate of iron, $FeTiO_3$, with 31.6 percent titanium, sometimes showing an irregular intergrowth of hematite or magnetite. Black; L black; S black; H 5 to 6; SG 4.5 to 5; very slightly magnetic.
Tests: BB infusible in OF, with fluxes reacts for iron in OF; with fluxes gives a brownish red color in RF; when this assay is treated with tin on charcoal it changes to violet red, showing the presence of titanium, which is significant. Filtered HCl solution with addition of tinfoil yields an excellent blue to violet color; not magnetic; its black streak distinguishes it from specular hematite.
Occurrence: As an accessory in many igneous rocks or as nearby segregations on the borders of the rock. Found in Norway; Switzerland; Chester, Massachusetts; Orange County, New York; a large deposit at Tahawus, New York; Litchfield, Connecticut; and Quebec. Alters to leucoxene, a dull, white opaque substance.

Rutile *(illus. pp. 145, 161)*
Tetragonal; crystals often prismatic, acicular prismatic; vertically striated; occasionally compact. A dioxide of titanium, TiO_2, with 60 percent titanium; iron in small amounts may be present. Color varies from colorless to dark if iron is present. L adamantine; S reddish brown to red; H 6 to 6.5; SG 4.2 to 4.3; brittle.
Tests: BB infusible; RF salt of phosphorous bead violet-colored; insoluble in acids; HCl solution of assay with Na_2CO_3, with added tinfoil gives a violet color if concentrated. Distinguished from tourmaline, vesuvianite, and augite by its adamantine luster, blowpipe reactions, higher SG.
Occurrence: An accessory in intermediate basic igneous rocks, gneisses, mica schists; in high-temperature veins and pegmatite dikes. Common as a resistant weather-released mineral, accumulated in placer sands. At many sites along the Appalachian Mountains such as Parksburg, Pennsylvania; Amherst and Nelson counties, Virginia; and Stony Point, North Carolina. Also in Magnet Cove, Arkansas. Fine large crystals from Graves Mountain, Georgia. An ore mineral of titanium.

TUNGSTEN (W)

The word tungsten means "heavy stone" in Swedish, and the mineral scheelite—calcium tungstate—was mistaken for cassiterite. However, a new element was isolated from this mineral and was named tungsten. Tungsten is a light gray metal closely related to molybdenum, chromium, and uranium. It is hard, malleable, and ductile and has great tensile strength. It is very heavy (SG 19.3, equal to that of gold). Tungsten has the highest melting point (3370° C) of any metal. It is used in alloys to improve steel properties, and in special alloys of extreme hardness that are used in high-speed tools. Tungsten carbide is exceeded in hardness only by boron carbide and by diamond; hence it is much used in machine tools and drilling bits. Tungsten is also used in filaments for incandescent lamps. Several minerals containing tungsten are of interest to collectors.

Wolframite

Monoclinic; crystals usually tabular, also prismatic and striated vertically, sometimes bladed, columnar, and granular. A tungstate of iron and manganese, $(Fe,Mn)WO_4$. This mineral is between the end members ferberite, $FeWO_4$, and hübnerite, $MnWO_4$, forming an isomorphous series. Brownish black; L submetallic; S brownish black; H 5 to 5.5; SG 7 to 7.5; C {010} perfect; brittle.

Opposite: Clockwise, from top left: Magnetite (an iron ore), limestone, and coke, used to make cast iron. *Above:* Chalcopyrite-pentlandite (top r.), ores of copper and nickel, used to produce monel metal, from which the screen, the metal bar, and the forceps are made.

Tests: Wolframite fuses easily to a magnetic globule; with salt of phosphorus in OF gives a clear reddish yellow bead; in RF becomes dark red; decomposed by aqua regia with separation of a yellow powder; HCl solution with zinc gives an intense blue, which fades on dilution.
Occurrence: These tungsten minerals are commonly found in granite and pegmatite dikes; commonly associated with cassiterite; also in veins with sulfide minerals. Some important localities are in Saxony, Bohemia, Romania, Spain, Portugal, Burma, Bolivia, and Cornwall, England. In the United States, in Trumbull, Connecticut; Black Hills, South Dakota; Nederland and Silverton, Colorado; and Gage, New Mexico.

Scheelite

Tetragonal, pyramidal; bipyramidal habit; also tabular, columnar, massive. A calcium tungstate, $CaWO_4$; molybdenum may be present. White to brownish, reddish; L subadamantine; S white; H 4.5 to 5; SG 5.9 to 6.1; C pyramidal; Fr conchoidal; brittle.
Tests: F 5 to a translucent glass; with salt of phosphorus OF green when hot. Strongly fluorescent in ultraviolet light. In HCl leaves a yellow powder; the solution when treated with tin and boiled yields a blue color. Its high SG distinguishes it from other nonmetallic minerals.
Occurrence: In pegmatite dikes and ore veins usually under hydrothermal conditions, also in contact metamorphic deposits. Some localities are Bohemia; Zinnwald, Saxony; Italy; Switzerland; Finland; Cornwall and Cumberland, England; and Mexico. In the United States can be found in Arizona; Trumbull, Connecticut; Custer, Lawrence, and Pennington counties, South Dakota; Mill City, Colorado; and Darwin, California. An important ore of tungsten.

URANIUM (U) *(illus. p. 120)*

Uranium oxide was discovered in 1789 in the mineral pitchblende, which up to that time was thought to be an ore of iron and zinc. The metal is brilliant and white, H about 5 but oxidizes to dark grayish black when exposed to moist air; SG 18.7, making it one of the heaviest of the metals. Uranium burns in air at 170° C. Its primary use is due to its natural radioactivity. For many years uranium ores were processed to obtain the much rarer and even more radioactive element, radium, for treatment of diseases. Today uranium is extremely valuable for its role in atomic power generation. Uranium compounds are also used in ceramics and glasses, glazes, photography, medicine, and in dyeing silk and wool. More than seventy-five minerals contain uranium as an essential element, but only a few occur in sufficient concentration to make their exploitation profitable. The bright colors of uranium minerals—chiefly reds, oranges, greens, and yellows—make them attractive to collectors.

Autunite

Orthorhombic; thin tabular crystals, also foliated, micaceous. A hydrous phosphate of uranium and calcium, $Ca(UO_2)_2P_2O_8 \cdot 8H_2O$. Lemon yellow; L pearly and subadamantine; S yellowish; H 2 to 2.5; SG 3.1; transparent to translucent.
Tests: BB same as for torbernite, except does not give copper.
Occurrence: Of secondary origin usually associated with uranium minerals. Autunite is often found in granite pegmatites without being closely related to earlier uranium minerals. In the dark it is easily detected with an ultraviolet lamp. A mineral found in association with other uranium minerals; outstanding specimens from Daybreak Mine, Mount Spokane, Washington. Also in Keystone and Custer districts, South Dakota; Colorado; Utah; California; New Mexico; and in New England.

Torbernite *(illus. p. 149)*

Orthorhombic; crystals usually square, thin plates; occasionally pyramidal, foliated, micaceous. A hydrous phosphate of uranium and copper, $Cu(UO_2)_2P_2O_8 \cdot 12H_2O$; arsenic may substitute for part of the phosphorus. Grass green; L pearly; S paler than color of mineral; H 2 to 2.5; SG 3.2; C {001} perfect; brittle.
Tests: CT yields water; BB colors flame green; salt of phosphorus yields a green bead; with Na_2CO_3 on charcoal gives a bead of copper. Soluble in nitric acid.
Occurrence: A late mineral in copper-uranium deposits. From localities in Saxony; Bohemia; Cornwall, England; South Australia; and Shinkolobwe, Zaire. In the United States from Chalk Mountain, near Spruce Pine, North Carolina.

Uraninite (Pitchblende)

Isometric; in cubes, octahedrons and cubo-octahedrons, nodular, dendritic, skeletonized crystals, arborescent, granular, massive. An oxide of uranium, UO_2, with variable amounts of lead, thorium, zirconium, radium, and metals of some rare earths and the gases nitrogen, helium, and argon. Many variations in composition are known. Black; L submetallic; S brownish black; H 5.5; SG 9 to 9.7.
Tests: BB infusible; with borax and salt of phosphorus gives a green bead in RF; soluble in nitric and sulfuric acids; nonmagnetic; strongly radioactive.
Occurrence: A primary constituent of granitic pegmatites; large masses in veins of silver, lead, copper, and other minerals. Some important localities are in Romania, Bohemia, Saxony, Norway, England, Zaire, and South Africa. In the United States at Grafton Center, New Hampshire; North Carolina; Gilpin County, Colorado; Marysvale district, Utah; the Ingersoll

Mine, Keystone, South Dakota; and Great Bear Lake. Also in Canadian Northwest Territories and western Ontario. An important ore of uranium.

VANADIUM (V)

Vanadium was discovered in 1830 and isolated in 1869. The metal is light gray in color and very hard, with a SG of 5.96 and a melting point at 1710° C. A small amount of vanadium in steel increases toughness, elasticity, and tensile strength. Such alloys are useful in automobiles, locomotives, and other machines containing parts subjected to severe stress. Vanadium compounds are also used in paints, printing fabrics, dyes, and medicine. Considerable vanadium is obtained from Venezuela and Mexico from flue soots accumulated by the burning of oil. More than fifty vanadium-bearing minerals have been described, but most are the products of weathering and oxidation of previously formed minerals. Many vanadium minerals are brightly colored and form lovely crystals prized by collectors.

Descloizite

Orthorhombic, dipyramidal; in subparallel growths, drusy crusts of intergrown crystals, massive granular. A basic vanadate of lead, copper, and zinc, $(Zn,Cu)_4 Pb_4(VO_4)4(OH)_4$. Brownish red, blackish brown; L greasy; H 3 to 3.5; SG 6.2. A secondary mineral of the oxidized zones, widely distributed in nonferrous mineral deposits. Can be found in Austria; Tsumeb, South-West Africa, as good crystals; Rhodesia; Katanga; England; Argentina; San Luis Potosí, Mexico; and in many localities in the western United States, for example: Silver Queen Mine, South Dakota; Lake Valley, New Mexico; and Schultz, near Oracle, Arizona. An ore mineral of vanadium.

Vanadinite *(illus. pp. 30, 149)*

Hexagonal, pyramidal; the most spectacular crystals of the vanadium minerals; prismatic with smooth faces, clean edges; terminated by either a base or by pyramids; sometimes barrel-shaped or globular in form, also skeletonized crystals. A vanadate of lead with lead chloride, $(PbCl)Pb_4(VO_4)_3$. Yellow to deep red; L adamantine; S white to yellowish; H 3; SG 6.7 to 7.1; brittle.
Tests: BB fuses easily on charcoal; in RF yields a lead globule; when completely oxidized in the OF, the residue yields a bright green bead in salt of phosphorus. In the OF it changes to light yellow.
Occurrence: A secondary mineral in the weathered or oxidized parts of primary lead and vanadium mineral deposits; not a common mineral. Some localities are in the Ural Mountains, Austria, Spain, Scotland, Morocco, Argentina, and Mexico, and throughout the southwestern United States, for example: in Arizona at the Apache Mine, Gila County; the Mammoth Mine, Tiger; the Red Cloud Mine, Yuma County; and the Old Yuma Mine, near Tucson. Also in Kelly, New Mexico. A source of lead and vanadium.

ZINC (Zn) *(illus. p. 168)*

Brass—the alloy of zinc and copper—was used long before iron was known, and is repeatedly mentioned in the Bible. Zinc became an important article of British commerce about 1800, and Germany was an important producer for about one hundred years. The first zinc production in the United States was about 1840 at Franklin, New Jersey. This locality is important to mineral collectors, because it has provided many new mineral species, some found nowhere else in the world.

Zinc is used extensively in alloys. Because of its resistance to weather and corrosion, it is used in coating sheet, nails, and other steel products against corrosion; this plating process is known as galvanizing. Zinc compounds are used in processing rubber, in ceramics, chemicals, agriculture, photoengraving plates for offset printing, and die casting. Zinc oxide is a popular white paint pigment.

Zinc is not found as native metal, but occurs as an essential element in about sixty minerals. Only a few are important as ores, but many zinc minerals are desirable to mineral collectors.

Hemimorphite *(illus. p. 97)*

Orthorhombic, hemimorphic; crystals thin, platy, prismatic, vertically striated, and attached by the same end, forming sheaves. When attached to reveal both ends, the crystals show the prominent hemimorphic development. A hydrous silicate of zinc, H_2ZnSiO_5. White with weak shades of color due to impurities; L vitreous to subadamantine; H 4.5 to 5.5; SG 3.4 to 3.5; C {110} perfect; brittle; transparent to translucent; strongly pyroelectric.
Tests: In CT decrepitates and gives off water; F 6; on charcoal with Na_2CO_3 gives a yellow coating, white when cool; moistened with cobalt solution gives a bright green color in RF; gelatinizes with acids. Blowpipe reactions and crystals separate it from similar minerals.
Occurrence: An oxidation product of zinc, throughout the world in weathered zinc deposits. In the United States found in Sterling Hill, New Jersey; Elkhorn, Montana; Leadville, Colorado; the Emma Mine, Salt Lake County, Utah; and Eureka, Oregon. An ore of zinc.

Smithsonite

Rhombohedral; rarely in crystals; frequently botryoidal, stalactitic, and as incrustations, granular and earthy. A carbonate of zinc, $ZnCO_3$; iron, cobalt, manganese, calcium, and magnesium may also be present. White, brownish, bluish, and green; H 5.5; SG 4.3 to 4.5; brittle.
Tests: BB infusible, yields char-

acteristic zinc flame and coating with Na_2CO_3; when moistened with cobalt solution and heated in OF gives a green color on cooling; soluble in HCl with effervescence. Distinguished from hemimorphite by its effervescence in acids.
Occurrence: An oxidation product of primary zinc minerals, found in many worldwide localities. The fine blue green specimens from Socorro County, New Mexico, are notable and provide an accent in mineral cabinets. Also found in the United States in Yellville, Arkansas; the Tin Mountain Mine, Custer County, South Dakota; Mineral Point, Wisconsin; and Granby, Missouri. An ore mineral of zinc.

Sphalerite (top l.), an ore of zinc, and chalcopyrite, a copper ore (top r.), with bars of zinc (bottom l.) and copper (bottom r.), which are combined to produce brass, as in the brass gears.

Sphalerite *(illus. p. 168)*

Isometric, tetrahedral; in tetrahedron, cube, dodecahedron, and other forms; crystals often distorted, massive, coarse- to fine-grained. A sulfide of zinc, ZnS. Nearly colorless, yellow, usually brown to black; L adamantine; H 3.5 to 4; SG 3.9 to 4.1; C perfect dodecahedral.
Tests: OT yields sulfurous fumes; BB on charcoal gives coating of zinc sulfide; in OF with cobalt solution the oxide coating gives a green color; dissolves in HCl with evolution of hydrogen sulfide. Distinguished by its crystal forms, cleavage, and blowpipe reactions.
Occurrence: Sphalerite is the most common mineral of zinc; a primary mineral in the upper and intermediate zones associated with galena and chalcopyrite, pyrite, calcite, barite, fluorite, and dolomite. Sphalerite is of very wide distribution in Bohemia, Romania, Switzerland, France, Spain, England, Scotland, Japan, and the Harz Mountains, Germany. In the United States in the area around Joplin, Missouri; in Roxbury, Connecticut; Franklin, New Jersey; the Wheatley Mine, Phoenixville, Pennsylvania; Mineral Point, Wisconsin; and Tiffin, Ohio; also in Canada and Mexico. The major zinc ore.

Willemite *(illus. p. 20)*

Trirhombohedral; in hexagonal prisms, also massive and in disseminated grains, sometimes fibrous. A zinc orthosilicate, Zn_2SiO_4; manganese often proxies for a considerable part of the zinc. White, greenish, yellowish brown to dark brown; resinous to vitreous; S none; H 5.5; SG 3.89 to 4.18; Fr conchoidal; brittle.
Tests: Fluoresces strongly in green color when irradiated with ultraviolet light. On charcoal in RF gives a yellow coating, white when cool; when moistened with cobalt solution and heated in OF becomes bright green. Soluble in HCl, leaving silica when evaporated.
Occurrence: Sparingly in zinc-ore deposits in Belgium, Algeria, Zaire, Zambia, South-West Africa, California, Utah, Arizona, New Mexico, and most notably at Franklin and Sterling Hill, New Jersey.

Zincite *(illus. p. 149)*

Hexagonal, hemimorphic; crystals rare; massive, granular. An oxide of zinc, ZnO, containing 80.3 percent zinc; the presence of a small amount of manganese may account for the color. Orange yellow to deep red; L adamantine; H 4 to 4.5; SG 5.43 to 5.7; C {0001} perfect; brittle.
Tests: BB infusible; with fluxes reacts for manganese; on charcoal gives zinc oxide coating, which when moistened with cobalt solution gives a green color in OF; soluble in acids. Distinguished by color and blowpipe reactions. Uncommon except at Franklin and Sterling Hill, New Jersey, and at the Dick Weber Mine, Saguache County, Colorado. An important ore mineral of zinc.

ZIRCONIUM (Zr)

Zirconium is used primarily in nuclear reactors, in chemical plants for corrosion-resistant material, and in photography in flash bulbs. Most zirconium is used in refractories, glazes, enamels, welding rods, and various alloys, especially special steels. The metal is hard and gray (SG 6.4). There are more than a dozen zirconium-bearing minerals, but only one is important both as an ore and to collectors.

Zircon

Tetragonal; usually in short prisms terminated by b pyramids; also as geniculated twins and in grains and irregular shapes. A silicate of zirconium, $ZrSiO_4$. Some varieties contain hafnium and sometimes rare earths. Colorless; yellowish to reddish brown; L adamantine; H 7.5; SG 4.2 to 4.8; Fr conchoidal; brittle; transparent to opaque.

Tests: BB infusible; heating makes profound changes in color; only fine powder is acted upon by acids. Decomposed by fusion with alkaline carbonates and bisulfates. Distinguished by its physical properties.

Occurrence: Zircon is a common accessory mineral of igneous rocks—especially the acid rocks—usually as small crystals; in pegmatites as large, well-formed individuals. Resistant to weathering and found commonly in weathered detritus from acid crystalline rocks. Found in the Ural and Ilmen Mountains, in France, Norway, Sweden, in the alluvial sands of Sri Lanka, and in Madagascar in large crystals. In the United States in Idaho and along the Appalachian Mountains, usually as small grains; in St. Peter's Dome, El Paso County, Colorado; Tin Mountain Mine, Custer County, South Dakota; Litchfield, Maine; Chesterfield, Massachusetts; Moriah, New York; Henderson County, North Carolina; and in eastern Ontario in very large crystals. An ore mineral of zirconium and a semiprecious gemstone.

THE ALKALI METALS

The three most common alkali metals—lithium, sodium, and potassium—are exceedingly active. They do not occur in the native state, are very unstable in the atmosphere, and react violently with water. Lithium occurs as an essential element in the minerals spodumene, lepidolite, amblygonite, and petalite. Sodium is present in the very abundant mineral halite. And potassium is present in acid feldspars, sylvite, and other minerals of salt deposits. The salts of these metals have widespread and divergent uses in the chemical industry, which affect our lives through many products. When needed, these metals are produced by electrolysis from their fused chlorides. They are used as reducing agents to produce such metals as titanium, zirconium, niobium, and tantalum from their compounds. Halite is an essential compound in the diet of humans and animals, while potassium is an essential element in the foods of plants.

USES OF NONMETALLIC MINERALS

In addition to the metallic minerals that supply the major metals used in industry, there are many nonmetallic minerals that are essential to the economy. Their uses are based entirely on their physical properties. In a few instances the chemical properties are basic requirements, in making cements or in the use of alkali, for example. The nonmetallics of similar properties or uses are listed here in groups.

The so-called rock-forming minerals are those minerals that are primary constituents of rocks. For example, most igneous rocks contain feldspar of some kind, frequently a mica, and various other members of certain major mineral groups. Some of these minerals have industrial uses, while others do not. The primary rock-forming mineral groups are the feldspars, amphiboles, pyroxenes, micas, garnets, and zeolites.

Amphiboles and Pyroxenes

(illus. p. 158)

Two of the principal groups of rock-forming minerals are the amphiboles and the pyroxenes. In the field the two groups can readily be separated on the basis of their cleavage angles, the minerals of both groups having good, although sometimes interrupted, prismatic cleavages.

The minerals of either group do not frequently occur in good crystal groups. They are usually massive and of a dark, unattractive color. Diopside occurs very rarely in gemmy crystals. Nephrite—a variety of the

From Sicily, superb crystals of sulfur, a mineral that is essential to modern industry. Its thousands of uses include the curing of rubber and the making of paper, matches, insecticides, and fertilizers.

tremolite-actinolite minerals—is rather common in some highly metamorphosed zones and is the source of gem and decorative materials. Jadeite—a pyroxene —is less abundant and more highly prized.

Feldspars *(illus. pp. 56, 84, 109)*

The feldspars are the most abundant minerals in the earth's crust. At or near the surface they are relatively easily disintegrated and decomposed by weathering. Their good basal cleavage promotes disintegration, which exposes many small pieces to weathering; these decompose into kaolinite and colloidal silica gel as well as soluble alkalies. All except the soluble portions enter into new rocks, the sedimentary rocks. The properties of the feldspars are much alike; they belong to either the monoclinic or triclinic crystal systems, have a hardness of six or close to this value, a perfect basal cleavage, and twinning according to several twin laws. The alkali feldspars—orthoclase, microcline, and albite—and some quartz, with the decomposition product kaolinite, have been used for many years as the chief ingredients in the manufacture of chinaware and industrial porcelain. Most of the feldspar was at one time quarried in Maine, New Hampshire, Connecticut, and North Carolina. In the pegmatites the feldspar grains were large enough to be hand-sorted in the quarry. The present efficient milling processes produce a high-grade feldspar concentrate from syenite rocks. Consequently, pegmatites are not actively worked at present as a source of feldspar for porcelain. This decline in the quarrying of feldspar has curtailed the production of many mineral specimens, which were frequently recovered by the miners or by mineral collectors from the quarry spoils.

Micas *(illus. p. 162)*

The micas are common rock-forming minerals and occur most abundantly in acid igneous rocks and metamorphic schists. Muscovite is the most common, as an accessory mineral in granites and syenites. It reaches its greatest development in granite pegmatites; sheets several feet across are frequently encountered near the wall rock. Phlogopite is sometimes asteriated. Both muscovite and phlogopite provide thin cleavage sheets, with a very high electrical resistance, which furnish excellent sheets for die-cut insulating forms for many electrical devices. The lithium mica—lepidolite—occurs chiefly in the soda-lithium rich granite pegmatites. Many rare minerals are associated with lepidolite. The mineral has only minor uses as a source of lithium and other alkali metals. Biotite occurs mainly in the basic igneous rocks, schists, and gneisses. The dielectric value is low, and consequently it has no economic value. Margarite, classed as a brittle mica, is associated with corundum in aluminum-rich metamorphic rocks and has no economic value.

Garnets *(illus. pp. 20, 36, 81, 181)*

The garnets present a fascinating group of minerals, many of which occur in excellent crystals from microscopic size to several feet in diameter, as at the Barton Garnet Mine near North Creek, New York. Even children are familiar with the word "garnet"; but they expect all specimens to be red, whereas some species are white, yellow, orange, or green. Garnets occur in both acid and basic igneous rocks, in sediments as placer deposits, and commonly in the metamorphic rocks, especially schists and gneisses. They are used as abrasives and as gems.

Zeolites *(illus. p. 210)*

The zeolite minerals are among the finest crystal groups; many of the species occur in excellent crystals. They are almost entirely within amygdaloidal cavities or open fissures in the gabbroic rocks, chiefly the diabases and basalts. They all contain water of crystallization, which some species lose at a low temperature or resorb with increased humidity. In addition to resorbing water, they may absorb other substances, such as ammonia and alcohol.

The zeolites are associated with a number of fine minerals —pectolite, apophyllite, datolite, prehnite, and calcite. Although zeolites have been considered secondary minerals, it is difficult to reconcile a secondary origin with the fact that they are commonly formed in basalt cavities. They are indeed derived from the late residual aqueous solutions given off by the solidifying magmas. The natural zeolites are too rare to be a factor in the economy, except as fine specimens for the collector.

Minerals Used for Abrasives

The four minerals most commonly used in the abrasive industry are diamond, quartz, garnet, and corundum. A few other hard minerals are suitable, but they are not sufficiently abundant. Diamond is bonded onto grinding wheels for shaping hard metals, such as the various carbide-tipped tools used for metal-cutting. Crushed quartz is sized and glued to tough paper for use in cleaning metals and smoothing wood; it is the least expensive.

There are more than six garnet minerals that are widely distributed chiefly in the metamorphic rocks; the iron-aluminum garnet—almandine—is the most suitable as an abrasive. Garnet has a conchoidal fracture. In the woodworking industries, garnet fragments are glued to paper for use in finishing wood; they eventually break under stress, and the fracture plane produces a new, sharp-edged cutting piece. Hence, fracturing provides sharp edges for continued use.

For many years corundum has been the leading mineral used in abrasive wheels, abrasive paper, and abrasive sticks. Several natural associates of corundum are magnetite, spinel, and diaspore. These minerals, together with corundum, are incorporated into the abrasive known as emery, which is used as loose grains, glued to paper, and in formed wheels and other convenient shapes.

The advent of the electric furnace has furnished a means of converting a number of mineral raw materials into superior abrasives. One of these, under very high temperature, converts coke and quartz into a very hard silicon carbide (SiC), which is used extensively for most abrasive purposes and has gradually replaced most natural abrasives.

Another common abrasive is derived from bauxite by purifying the alumina and fusing it in the electric furnace, which converts it into synthetic corundum, another artificial material widely used as an abrasive.

Clay Minerals and Products

(illus. pp. 153, 158)

Clay is the one mineral product that touches nearly everyone every day in one way or another. The manufacture of products using clay is a multimillion-dollar industry that starts with an abundant and low-cost natural product. Because of the haphazard conditions under which clays originate from a great diversity of primary products, their compositions vary widely. However, the great range of uses of clay products can accommodate the many different types. Most clays form from the alteration products of weathered feldspars. Little notice is taken of them by the average mineral collector, although more than a dozen clay minerals have been described. They are mostly of a white or cream color.

The principal clay mineral is kaolinite, which occurs in nearly pure beds that are widely distributed. In various amounts, it also forms a part of most other clay deposits. The finest and purest white clays are the basic ingredients in fine china and art objects. Clays of varying compositions and physical properties are used in a wide variety of products—porcelain, floor tile, electrical fixture supports, insulators, fire bricks, stoneware, terra-cottas, bricks, and drain pipes. Most of the paper we use contains either clay or talc. The oils we use for cooking may have been purified by clay filter presses.

A few alumino-silicates are used in making the most refractory porcelains. These are not clays; however, like feldspar and other aluminum silicates, they can be made into superior porcelain products. Among these are spark plugs for internal-combustion engines, small crucibles used in chemical laboratories, and fire brick used in high-temperature furnaces and kilns.

An interesting mineral is sepiolite, microcrystalline ($H_4Mg_2Si_3O_{10}$; H 2 to 2.5; SG 2; white and tints of other colors). This dry, porous material floats on water. It is an alteration product of serpentine and magnesite and occurs in Asia Minor in stratified deposits. In commerce the material is known as meerschaum and is used for making fine quality pipes for tobacco smokers.

Adularia (Switzerland), a variety of orthoclase feldspar, one of the chief ingredients in the manufacture of fine china and porcelain.

8. *Gems and Ornamental Stones*

The first use of gemstones must go far back into the age of primitive human beings. Although their main activity was to find enough food to feed themselves, they must at times have been intrigued with the natural objects around them. Eventually, bright water-tumbled pebbles attracted their attention. The prettiest stones would naturally have become objects of barter; and finally that dominating attribute of humans, the pride of possession, was developed, and it has been with us ever since.

Some tribespeople always wear on their persons their entire collection of valuable materials, because that is the safest way to protect them. Most individuals of so-called civilized countries today wear relatively few pieces of jewelry. Some of these pieces and the manner in which they are worn convey messages about their owners. You may be wearing a piece of jewelry in mourning for a relative; as an announcement that you are engaged or married, or are a member of some school or religious sect; or to indicate your status in society. Some great jewels represent the powers of governments and the officials of such governments. Nearly every person in the affluent societies of the world prizes at least one trinket bearing a piece of the earth.

If we were to ask anyone wearing a gemstone why he or she wore it, the answer might be that its color was pleasing, that it was a gift, or that it was a lucky stone. In the not too distant past, stones were often worn or carried as talismans, for they were believed to have the extraordinary power of repelling evils, diseases, or other effects of witchcraft. With increased education and understanding of natural phenomena, the superstitious belief in the power of stones has vanished, and they are now worn for what they are—beautiful pieces of the earth's crust. One of the more common reasons for wearing distinctive stones is fashion; another is the custom of wearing a stone designated for the month of one's birth.

When we consider gems, one of our first thoughts usually concerns the price. We have become accustomed to the high prices of gems and therefore consider them to be "precious." In antiquity the so-called precious gems included diamond, emerald, ruby, sapphire, pearl, and sometimes opal. Other stones were called "semiprecious," and the distinction was based primarily on scarcity and value. Today, however, some rare varieties of stones once considered semiprecious can be very expensive. Some rare green garnets, for example, may cost more than $1,000 per carat, which is much higher than the cost of a poor diamond. So the terms precious and semiprecious really no longer have much meaning and should be abandoned. We may consider all gems as precious if they are desirable. Many opaque gems, such as the inexpensive chalcedonies, agates, and jaspers, would not usually be called precious; they can be considered decorative stones.

The actual value of a gem is determined by many factors. Among them are color, hardness, durability, scarcity, freedom from imperfections (such as inclusions in faceted gems), cutting quality, and the fashion of the times. The scarcity and physical attributes, such as hardness, are easily evaluated; the more subjective qualities—such as intensity and shade of color and the style of cutting—are generally lumped together into a general term: beauty. The qualities that make a gem "beautiful" are subject to the whims of fashion and the society in which the gem is worn.

GEMS

Diamond

Only about 20 percent of the diamonds recovered from the earth are used in the gem trade, while the far larger 80 percent are of the greatest importance in industrial operations. Nevertheless, the total value greatly exceeds that of those used in industry. To be of superior gem quality, a stone must be hard and have a high resistance to abrasion; it must be rare, though plentiful enough to supply a substantial market; and it must be of superior beauty, because of either its color or the play of colors caused by the dispersion of light passing through the cut stone. Although the diamond has a perfect octahedral cleavage, its tenacity between cleavage planes prevents frequent fractures. Diamond crystals are frequently in good octahedrons and sometimes in cubes. Many crystals are rounded or of a tabular shape.

Although most of the gem-quality dia-

Preceding pages: Cut tourmaline gems from Dunton Mine, Newry, Maine.
Opposite: Topaz crystal (uncut), from Tanokamiyama, Japan.

monds recovered average less than a few carats, a number of large stones have been found. The largest is the Cullinan Diamond, a white diamond of 3,106 carats, found in South Africa in 1905. It is three times heavier than the next largest known diamond, the white Excelsior, 995.20 carats, found in South Africa in 1893.

The color of gem diamonds varies from yellowish to tints of the other colors of the spectrum. The usual color of first-quality diamonds is called blue white, while the off-color stones of much lower value are of a yellowish tint. However, the deep-colored stones are highly regarded and sell at an advanced price. One of the finest large yellow or golden-colored diamonds is the 128.51-carat gem owned by the Tiffany Company of New York. This stone was found in the Kimberley Mine in 1878. The 44-carat Hope Diamond is deep blue and may be seen in the United States National Museum in Washington, D.C. One of the most famous diamonds of all, the Koh-i-noor, has the longest known history; it has been traced as far back as 1304 but is believed to have been found some centuries earlier. Intrigue, violence, and war were a part of the Koh-i-noor's existence until 1849, when it was seized by the British East India Company. A new cutting, undertaken at the order of Queen Victoria, failed to give the stone its anticipated brilliance, and the diamond subsequently lost all resemblance to the original Koh-i-noor, a great historical loss.

The color of some brownish diamonds can be improved by exposing them to a high energy beam in a cyclotron or by placing them in an atomic pile. Such color changes improve the appearance of the gems and increase their intrinsic value, unless one objects to treated gems.

Diamonds are valued chiefly for their play of color or fire, which is caused by their high index of refraction (2.42) and their dispersion (0.058). Only a few minerals have a higher refractive index or dispersion. Diamonds occur in very specialized basic rocks in the necks of volcanoes. The outstanding locality in the

United States is Murfreesboro, Arkansas; they are also found in Illinois and West Virginia.

Gems that closely resemble the diamond in appearance are colorless varieties of zircon, rutile, anatase, cassiterite, and titanite. Most of these minerals are normally colored and consequently are easily distinguished from diamond.

Ruby and Sapphire

These two gems, one red, the other blue, are color varieties of the mineral corundum. Other colors of this mineral also occur, including green (green corundum), yellow (golden corundum), and violet. In general corundum is translucent to opaque, and gray, bluish, or brown to black in color. Crystals are usually hexagonal prisms. The indices of refraction are $o = 1.768$ and $e = 1.760$, giving a low birefringence. The dispersion is 0.018, which is low; hence, there is very little play of color. Due to the dichoric property of corundum, faceted stones should be cut with the table parallel to the base for the best color.

The composition is somewhat variable; small amounts of impurities are often present, and it is one or more of these that accounts for the various colors. The presence of chromium in small amounts is believed to give the red color, titanium the blue, and iron the yellow.

Oriented acicular crystals of other minerals may be present. Such fine crystals give a six-rayed asteriated effect when viewed perpendicular to the base, and when cut *en cabochon* will display a strong six-rayed star, producing the fine star ruby or star sapphire.

The word "sapphire" has come to designate the dark blue color of the mineral; however, because of its direct association with the extremely valuable blue transparent variety and the translucent star sapphire, the term has also been used to designate varieties of corundum in the other colors, except the red. For example, a red corundum in weaker color than ruby is called a pink sapphire; others are known as yellow, green, and violet sapphires. These stones have also been called "oriental emerald," "oriental topaz," and "oriental amethyst," but such terms are misnomers. Furthermore, they

Opposite: Uncut tourmaline crystals from Dunton Mine, Newry, Maine. *Left:* Diamond crystal in kimberlite matrix, from South Africa. *Above:* The DeLong Ruby (100 carats), a star ruby, from Burma, now in the American Museum of Natural History, New York.

are misleading, and since they may confuse the prospective purchaser about the exact nature of the stone, they should be avoided. These gems can be accurately described as pink, yellow, or green corundum.

The gem corundum crystals originate chiefly in metamorphic basic rocks that contain aluminum-rich minerals, in impure marbles, gneisses, and schists, and in some aluminum-rich igneous rocks. Some of the associated minerals are spinel, periclase, tourmaline, kyanite, magnetite, chlorite, and nephelinite.

The individual crystals or grains of gem-variety materials are not abundant in the rocks in which they originate. During the many millions of years that these source rocks have been exposed at the surface, weathering processes have freed the corundum gem materials from the rock, along with many other resistant minerals. These substances accumulate in streams and are deposited with gravels in valleys. The higher specific gravity of corundum (like that of most other gem materials) has caused it gradually to move downward to the base of the gravel layers, where it has been concentrated. Mining operations are primitive. To reach the gem-bearing gravels, open pits are sometimes developed. Often, crude shafts are sunk to the base of the gravel layer, and the gravels accessible from the bottom of the shaft are raised to the surface. This material is washed to reveal the gem minerals. Much water is encountered in some gravels, which increases the difficulty and cost of obtaining the gems.

The area in the vicinity of Mogok, in upper Burma, is the source of rubies with the best characteristic blood-red color. In this area, the source rock is a highly metamorphosed limestone that is coarsely crystalline. The gem materials have weathered from the rock and have accumulated in sands and gravels. Associated with the ruby are spinel, tourmaline, sapphire, and zircon. Thailand is also a producer of rubies and sapphires from gem gravels, which likewise contain zircon and red spinel.

From northern India, in the district of Kashmir, very fine sapphires have been obtained. A number of excellent gemstones, including rubies and sapphires, have been derived from the gravels of Sri Lanka (Ceylon). Other countries of southeastern Asia have also yielded some corundum gem minerals. A few fine rubies and sapphires have been found in the United States in metamorphic rocks: gem gravels in the Cowee Valley of North Carolina have produced a few fine rubies and sapphires; and gem sapphires have been mined from an igneous dike in Yogo Gulch, near Helena, Montana.

Opposite: Aquamarine cut from beryl found lying loose at Stoneham, Me., now in the Field Museum, Chicago. *Above:* Quartz crystals from Mt. Ida, Arkansas, with cut quartz brilliant and star-shaped quartz gems.

Inferior ruby and sapphire crystals were formerly cut and used as jewels or bearings in fine watches and delicate scientific instruments. Following the discovery of the art of synthesizing ruby and sapphire by Auguste Verneuil in 1902, synthetic transparent corundum became the basic material for instrument bearings. This development has placed on the market vast quantities of synthetic rubies and sapphires but has not destroyed the market for the natural stones. Rubies of a superior color and flawless consistency are extremely valuable and highly prized. The environments under which the natural and synthetic corundum are formed are very different, of course, and each material carries with it some effects inherited from its formative process. In the natural sapphire, we find a zoning of straight lines parallel to a crystal face, while in some synthetic ruby the zoning lines, due to accretion, are curved; a careful examination of the synthetic may reveal tension cracks and carbon inclusions. The inclusions found in the natural corundum gems would be minute specks of the associated minerals. Improved manufacturing has given a cleaner synthetic product.

Beryl

Beryl crystals are occasionally unusually large and weigh many tons; yet within these large masses there is very rarely even a small bit of gem material, for the conditions of crystallization are not often conducive to the development of clear, transparent crystals or grains.

Beryl is slightly harder than quartz. It is relatively light in weight, giving a larger gem per carat than diamond. Gem beryl crystals occur in a wide variety of colors, depending on the slight admixture of other elements in the crystal lattice, such as chromium, lithium, or cesium. The colors range throughout the spectrum, with the grass green variety—emerald—ranking high among gems in value. Emeralds of excellent quality are rare and very costly, while those of inferior quality, although not abundant, are readily available. The color of emerald varies greatly, from bright grass green to dull, translucent green; there is also a light green, which is more properly classed as green beryl. The outstanding United States locality for emeralds is Stony Point, North Carolina.

Opposite top left: Emerald crystals—a beryl gemstone—from Chivor Mine, Almeida, Colombia. *Opposite top right:* Golden topaz crystals from Minas Gerais, Brazil. *Opposite bottom:* Crystals of watermelon tourmaline from Newry, Maine, cut across to show the "watermelon" configuration. *Top:* Brooch made with materials collected by the naturalist George R. Howe in Oxford County, Maine: gold panned from the Swift River, amethyst from Pleasant Mountain, and pearls from freshwater clams. *Above:* Cut spessartine garnet gem.

Aquamarine is the blue or blue green variety of beryl. In ancient times, the most highly prized color was "the color of the sea," from which the name arises. Today, however, the bluer gems are often more highly prized. Unlike emerald, aquamarine crystals of gem quality weighing many pounds are found occasionally, and large transparent gems with no internal imperfections can be cut. Sometimes pale blue-green aquamarine of inferior color may be heat-treated, a process that eliminates much of the green and leaves a bluer gem.

The yellow and orange variety is called golden beryl, a suitable name for this beautiful and brilliant golden yellow gem. Nearer the red end of the spectrum are the pink or rose red beryl gems, known as morganite. These gems are found in the United States in Pala, California, and in Haddam, Connecticut. Another variety is the clear, colorless goshenite, which received its name from the locality in western Massachusetts where it was first discovered. This variety is associated with lithium-cesium minerals and contains a small amount of cesium. Goshenite is not uncommon in small amounts in the sodium-lithium-cesium pegmatites; some localities are Maine, Connecticut, California, and Goshen, Massachusetts.

Emeralds of superior color, free from flaws and inclusions, are extremely rare. Beryl has a strong tendency to engulf or incorporate other minerals within the crystals as inclusions. This sort of development is not uncommon in pegmatites. When beryls develop in schists, inclusions of schist minerals are common; hence, the emeralds originating in schists are usually badly flawed and abound in inclusions, which place them in a low price range.

The best emeralds are obtained from mines in the Muzo and Chivor areas in Colombia. The lithology of these mines consists of crystalline calcareous rocks, in which the emerald crystals—usually of small size—occur in calcite-lined cavities. In order to produce the emerald variety of beryl, the element chromium must be present and incorporated in the crystal lattice. This element is practically absent from the acid igneous rocks in which the vast amounts of beryl occur; on the other hand, beryl is extremely rare in the basic rocks where chromium is common. Hence we have the reason for the dearth of emerald crystals. It is only by rare coincidence that the geological conditions are suitable for the development of this gem.

The other varieties of gem beryl are found principally in sodium-lithium-rich acid pegmatites, which have a worldwide distribution. In these occurrences the gem beryl crystals are frequently found in larger and clearer crystals. One of the outstanding aquamarines, obtained at Morambaya, Minas Gerais, Brazil, weighed 243 pounds and was so clear that newsprint could be read through its depth of nineteen inches. The crystal sold for $25,000, but would be worth much more today. Gem beryl also occurs in Madagascar, the Ural Mountains, the Island of Elba, India, Sri Lanka, Maine (Middletown and Portland), Connecticut, North Carolina, New Hampshire (Acworth), Colorado (Mount Antero), and Massachusetts. Golden beryl is found in Sri Lanka, Brazil, Maine, and Connecticut.

Modern chemical techniques have developed processes to produce the gem variety of emerald synthetically. Synthetic blue-green spinel has been produced and misrepresented as aquamarine. Glasses and combinations of glass with natural gem material have been used to mislead purchasers into acquiring inferior gems.

Topaz

In early times the word "topaz" was used to designate yellow to orange stones, before the name "topaz" was scientifically attached to the naturally occurring orthorhombic compound. Apparently for economic reasons, the trade in some instances still clings to the old, unscientific descriptive practice, to the confusion and possible deception of many unknowledgeable purchasers of yellow gems. Terms like "oriental topaz," "Scotch topaz," "false topaz," "Madeira topaz," and "smoky topaz" should be avoided.

Topaz has a hardness of 8, a specific gravity of 3.4 to 3.6, and a perfect and relatively easy basal cleavage. The luster is vitreous. The rather weak colors of the transparent varieties range across the spectrum; light-colored, dense opaque crystals and masses are encountered in pegmatites and acid igneous rocks. The index of refraction is 1.62 to 1.63, and the dispersion is weak (0.014); hence, the cut stones lack fire but are attractive in either brilliant or step cuts.

Brazil is an important producer of both colorless and colored topaz; large crystals have been mined at Ouro Preto. The largest of the colorless crystals weighs six hundred pounds and is now in the American Museum of Natural History in New York. Other large topaz crystals from Brazil are in the National Museum in Washington and at Harvard University. Some of the common minerals associated with topaz are tourmaline, quartz, fluorite, apatite, beryl, cassiterite, and wolframite. Green and blue topazes have been mined near Ekaterinburg, Miask, and on the Sanarka River in the Soviet Union, as well as in Scotland, England, and Sri Lanka. Japan has produced some fine crystals. In the United States, the Thomas Range in Utah has supplied many small, well-developed clear crystals. Topaz also occurs in San Diego County, California; the Pikes Peak region, Colorado; Lord's Hill, Maine; Baldface Mountain, New Hampshire; and Trumbull, Connecticut.

Tourmaline

Tourmaline is one of the most fascinating and beautiful gem minerals. Crystallizing in the hexagonal system, it is pyroelectric and piezoelectric and consequently is employed in many sophisticated pressure and electrical devices. Its hardness is 7 to 7.5, which is a little low for abrasion resistance in ring stones or other uses subjected to severe wear. It is hard enough for brooches, pendants, and pins, however. Tourmaline has a specific gravity of 3 to 3.2, a conchoidal fracture, and a vitreous luster. The color varies from colorless to black and may include almost any hue of the visible spectrum. Some crystals show a marked zoning, usually parallel to the trigonal prism, or a dark blue or red interior with a green exterior shell. In other cases the base may be red with a green covering and green terminal sections. The last crystals to develop after the red phase has ceased to be deposited are usually entirely green. Tourmaline furnishes many beautiful gems, some of the strong green crystals yielding gems closely resembling the color of emeralds.

Locality names have been used to describe some of the varieties of gem tourmalines, occasionally in combination with the names of other minerals, such as "Brazilian emerald" for the green variety. Such names are misleading and completely erroneous. The red variety should be known as rubellite; the yellow, yellow tourmaline; the green variety, tourmaline; the blue or violet, indicolite; and the brown, dravite.

The black variety is known as schorl and is a rock-forming mineral in many schists and gneisses, especially in the vicinity of pegmatites and in the body of many potassium-rich pegmatites. The colored varieties are usually associated with the sodium-lithium-cesium pegmatites and are found in numerous localities; the leading ones are Brazil, Madagascar, South-West Africa, the Island of Elba, Sri Lanka, India, Burma, and Siberia. In the United States, colored tourmalines are found at a number of localities in Maine, Massachusetts, Connecticut, New York, Pennsylvania, and California.

The tourmaline crystals of the Dunton Gem Mine, Newry, Maine, are noted for their red centers surrounded by a green shell. These colorful crystals were named "watermelon tourmaline" by the late George R. Howe, a noted naturalist of Norway, Maine. Tourmaline is also found in Paris, Maine; Haddam, Connecticut; Gouverneur, New York; and Pala, California.

Garnet

Garnet is one of the best-known gems; however, the word pertains not just to one gem, but to a group of minerals with similar properties and appearances. Garnets are isometric and occur chiefly as dodecahedrons and tetragonal trisoctahedrons. The indices of refractions are variable, from 1.70 to 1.94. The dispersion is close to 0.025 for most varieties, but andradite has a dispersion of 0.057, which is close to that of diamond, and light-colored stones of this species have exceptional fire.

Grossular garnet occurs chiefly in metamorphosed limestones, and most of the gems come from the gem gravels of Sri Lanka. In the United States it is found in Minot, Maine; Warren, New Hampshire; Eden Mills, Vermont; and Inyo, Riverside, and San Diego counties, California. Pyrope is the garnet most frequently cut into gems, probably because of its fine, lively red color, and it is the variety that occurs in the diamond mines of South Africa. It occurs in the United States in San Juan County, Utah; Apache County, Arizona; Murfreesboro, Arkansas; and Macon County, North Carolina. Spessartine, the brown variety, lacks the popularity of the red garnets; it occurs in granites and metamorphic rocks rich in manganese. Some localities are San Diego County, California; Ely, Nevada; Rutherford Mine, Amelia County, Virginia; New Mexico; and North Carolina. Almandine, the iron-aluminum garnet, was known as "carbuncle" in ancient times and is deep red to violet red. In the United States, it is found in Rangel, Alaska; Emerald Creek district, Benewah County, Idaho; Black Hills, South Dakota; Roxbury, Connecticut; and Barton Mine, North Creek, New York. Andradite, the calcium-iron garnet, is variable in composition, has a high dispersion, and appears in many colors because of its variation in composition. When the color is weak, the fire is exceptionally good in faceted stones. It is found in the United States in Garnet Hill, Calaveras County, California; Iron Hill, Gunnison County, Colorado; Magnet Cove, Arkansas; and Flint and Sterling Hill, New Jersey. The chromium garnet, uvarovite, usually occurs in crystals too small for cut gems; however, a few of the larger gems closely resemble the emerald in appearance. They are found in the United States in northern California and in Riddle, Oregon.

Many descriptive names have been given to garnets to promote sales, such as "Cape ruby," "Arizona ruby," "Uralian emerald," and "Transvaal jade." All of these are false and misleading and should be avoided.

GEMS OF THE SILICA MINERALS

The silica minerals occur in two principal groups—the well-crystallized quartz varieties, and the cryptocrystalline chalcedonic varieties.

Quartz

Probably the most abundant mineral to occur in almost perfect crystals is quartz. In some occurrences the well-proportioned crystals are completely covered with smooth, glistening faces; this high degree of perfection is aston-

ishing and greatly admired by all collectors. Quartz also occurs in large crystals of exceptional perfection and clarity, as shown by the large, highly polished sphere in the National Museum in Washington, D.C. Although many quartz crystals are clear and free from internal imperfections, they are seldom used for faceted gems because of the low mean index of refraction (1.55), low double refraction (0.009), and low dispersion (0.013). Quartz has been used frequently for art objects. It is often referred to as rock crystal, but such terms as "Herkimer diamond," "Alaska diamond," and "Cornish diamond" are improper and deceptive.

Transparent quartz is found in quantity in Middleville and Little Falls, Herkimer County, and Ellenville, New York; Hot Springs, Arkansas; Pala, California; Chestnut Hill, North Carolina; and many other localities.

More frequently cut into attractive, faceted gems are the smoky varieties of quartz, called "cairngorm" in the lighter colors and "morion" when nearly black. The name "Scotch topaz" for this material is erroneous. It occurs in the United States in Auburn, Maine; Alexander and Lincoln counties, North Carolina; Pikes Peak, Colorado; and Butte, Montana.

Amethyst, the purple variety, is widely distributed. The quality of the stone varies greatly, from fine-colored specimens to dull translucent material; the color is distributed through the crystals either irregularly or, occasionally, in zones. Large crusts of deeply colored amethyst carrying gem-quality material occur in Brazil and Uruguay. Sri Lanka, India, Iran, Mexico, Pennsylvania, and North Carolina have also produced good amethysts, as have several localities in Maine, principally Deer Hill and Pleasant Mountain, as well as Keweenaw Peninsula, Michigan, and Coos County, New Hampshire.

The pink variety of quartz, called rose quartz, occurs chiefly in pegmatites, usually in granular masses varying in size from small pieces to those weighing several tons. Some rose quartz is asteriated and makes excellent beads for necklaces or brooches. The material is used in sculptures by lapidaries of Southeast Asia. Rose quartz rarely occurs in crystals, and the scarcity of good cabinet specimens results in a high price for this variety. Massive rose

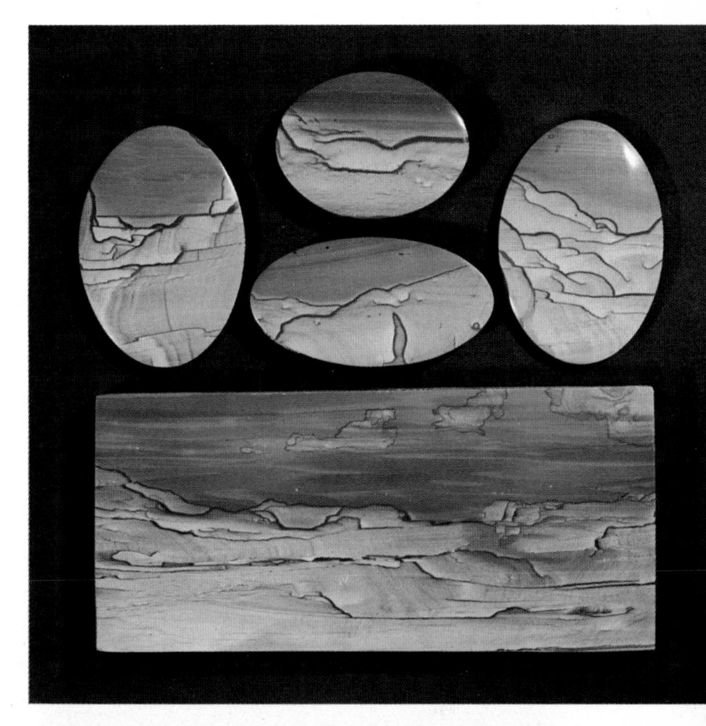

Opposite top: Fortification agate, from Custer State Park, South Dakota. *Opposite center:* Amethyst (top) changed to citrine (bottom) by heat treating. *Opposite bottom:* Excellent examples of moss agate. *Top:* Picture jasper, cut in cabochons and rectangle, from Succor Creek, Malheur County, Oregon. *Above:* Petrified wood, showing original cell structure, from southwestern U.S.

quartz is found in France, Bavaria, Madagascar, Japan, and Brazil, and in the United States in Maine and South Dakota. Crystals are found in Maine and Brazil. Specific localities are Scott Rose Quartz Quarry, Custer County, South Dakota; Paris, Maine; Lagrange, Georgia; and Pala, California.

Citrine, the yellow variety, is usually found as crystals and in the past has been sold fraudulently as topaz. Poor grades of amethyst can be changed to citrine by heat treatment. Citrine is found in the United States in the Strickland Quarry, Portland, Connecticut, and in Bedford, New York.

Some quartzites contain minute, scalelike specks of hematite, goethite, or chlorite, which give the stone a flickering yellow, brown, green, or red color when rotated. This rock is known as aventurine quartz.

Under certain conditions, fissure cavities develop a multitude of acicular, prismatic rutile crystals, which grow out from the walls in a crisscross fashion. Following the crystallization of the rutile, large quartz crystals grow in the same space and completely enclose the rutile, resulting in an attractive material known as rutilated quartz or sagenite. This material is found in Stony Point, North Carolina; Sycamore Mills, Pennsylvania; and Lagrange, Georgia. It is cut *en cabochon* or is used as fine cabinet specimens. Hornblende, actinolite, and other slender, prismatic minerals also occur as inclusions in quartz.

Chalcedony Materials

Chalcedony is a cryptocrystalline variety of quartz that occurs in variously shaped masses or crusts in rock cavities and sedimentary rocks. It is the source of a number of gem materials, which take their names from the kind and distribution of the included substances. The common impurities are oxides of iron, nickel, aluminum, manganese, and weathering products of the basic minerals.

The best known and undoubtedly most distinct of the chalcedony gem materials is agate, which consists of concentric narrow bands and is found in many localities. The banding is sometimes marked by a strong contrast between the red and the white colors, which is produced by the alternating precipitation of chalcedony and the impure material, which usually contains hematite. Some agates are artificially colored with dyes. When the

Opposite: From Cape Province, South Africa, uncut tiger's-eye, a chalcedony that displays change of colors. *Top:* A slab of poppy jasper. *Above:* Jasper (dark) and chalcedony (light), used for cabochon gems.

Above: Squash blossom necklace and earrings, silver set with turquoise from the Shoshone Mine, Austin, Nevada. *Opposite top:* Slices of variscite from Fairfield, Utah. *Opposite center:* Jade brooch stone.

banding forms an angular design, the agate is said to have fortification banding. Fortification agate is found in the United States in Fairburn, South Dakota; Braen's Quarry, Hawthorne, New Jersey; Muscatine, Iowa; and Cook County, Minnesota. Agates originate through either igneous or sedimentary processes.

Chalcedony results from the consolidation of colloidal silica. While in the gelatinous state, it may be penetrated by ramifying growths of other substances, which produce characteristic designs. Of these substances, the oxides of manganese and some chlorite and amphibole minerals develop mosslike, dendritic, or acicular structures, about which faint banding develops during consolidation. These structures are known as moss agates and are found in Chestnut Hill, North Carolina; William and McKenzie counties, North Dakota; Douglas County, Nebraska; and Wyoming. Sometimes picturesque designs develop, and when the stones are cut across they give the impression of a silhouetted landscape. Such scenic designs are known as landscape agates.

When red iron compounds permeate the colloidal silica, the resulting hardened mass, called carnelian, is red. Carnelian is found in Stirling, New Jersey; Van Horn, Texas; Pacific County, Washington; and Clatsop County, Oregon. Nickel compounds in the colloidal silica lend a green color to the hardened material, which is called chrysoprase and is found in Venice Hill Mines, Exeter, California. This stone is hard, takes a good polish, and makes splendid green cabochon stones for rings, pins, and brooches.

The green cryptocrystalline varieties of quartz are known as prase. It is found in South Mountain, Redington, Pennsylvania. The color varies from light to dark green and may be bright to dull depending on the coloring agents, which are weathering products of the basic minerals. Other sources are China and India. Bloodstone, found in Los Angeles County, California, and elsewhere, is a dark green prase with bright red splashes of jasper that look like spots of blood. Christian churches have favored this stone for use in sacred objects, since the jasper is thought to represent the blood of martyrs. The stone, used also in signet rings, is found in India, the Hebrides, and Siberia.

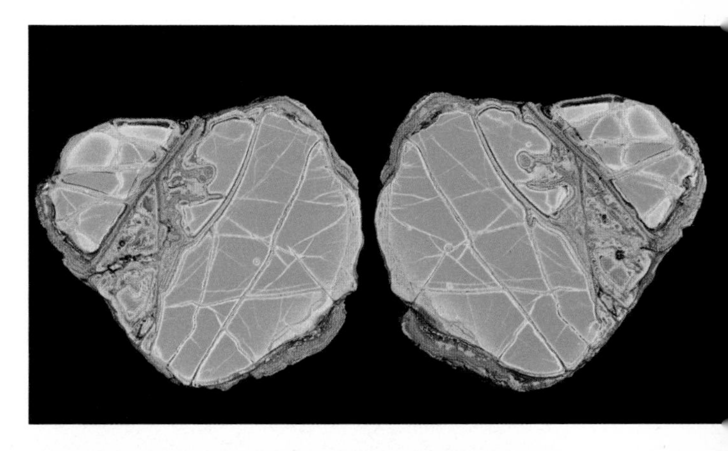

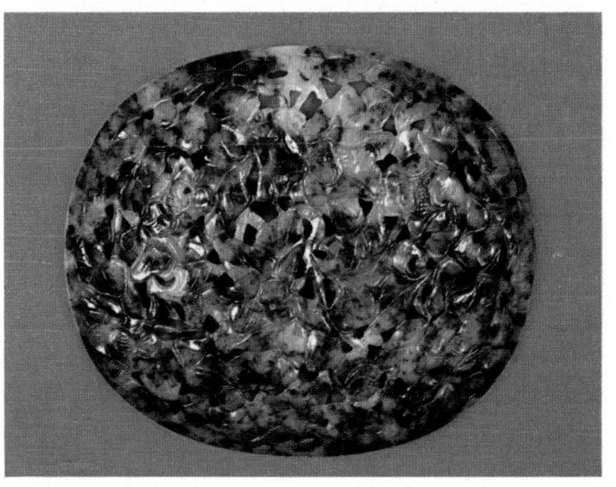

Jaspers

When they weather, most silicate minerals yield quantities of colloidal silica. This is carried by streams to a settling basin, where the colloidal products are deposited as siliceous layers or lenses and form distinct concretionary masses, in irregular shapes and even sheets. Along with the colloidal silica are colloidal solutions of other elements from the original rock minerals, chiefly iron and possibly some aluminum. All this material, which is in various colors, is incorporated in the deposits to constitute a plastic mass.

During the consolidation process, the gradual loss of water creates differential pressures in the material, producing the phenomenon of plastic flow; the end result in the hardened jasper resembles the effect of marbled cake. The plastic flow and differential settling creates rocks that appear to contain colorful pictures; these intriguing and often beautiful stones are popularly cut in slices and offered for sale in rock shops. One of the most striking picture rocks, from the vicinity of Succor Creek, Oregon, displays what looks like a landscape topped by a blue sky with clouds. Other localities are Chittenden County, Vermont; Spies Church, Alsace, Pennsylvania; Woods County, Oklahoma; San Juan County, New Mexico; Maricopa County, Arizona; Douglas County, Oregon; and Santa Clara County, California. Jaspers also provide colorful cabochons for jewelry and other ornaments.

Onyx is a banded stone consisting of straight, black and white parallel lines, found at Priday Ranch, Madras, Oregon, and other localities; in sardonyx the parallel layers consist of various colors, including red carnelian and white chalcedony. This is the material that is used for the classic Italian cameos. A banded calcareous marble from Tecali, Puebla, Mexico, is known as Mexican onyx and is used for art objects in Mexico and the United States.

Petrified wood—from Arizona; Eden Valley, Wyoming; Palo Duro Canyon, Texas; Kittitas County, Washington; and elsewhere—is essentially a colorful jasper derived from silicified wood in which the outlines of the wood cells are still microscopically visible. It is used for cabochons and other decorative objects.

Tiger's-eye is the popular name for a type of silicified mineral, crocidolite. The silicification produces a hard chalcedony of various colors, from silky golden yellow to shades of blue. The material takes a good polish and is cut into slabs for ornamental uses and into cabochons for jewelry. The main source of supply is in Cape Province, South Africa; it is rarely found in the United States.

Opal

Opal forms from the agglomeration and dessication of colloidal silica, which has accumulated in the narrow cracks and spaces of sedimentary rocks in arid regions, as in Australia; it also forms as small globules of colloidal silica, immiscible fractions in the magmas of some acid igneous rocks, as at Queretaro, Mexico. Opal is relatively soft; it has a hardness of 5.5 to 6.5 on the Mohs scale and a specific gravity of 1.95 to 2.3; it is amorphous, and the index of refraction is 1.44 to 1.46. These low values would be enough to deny the material status as a gemstone, were it not for the opal's superior fire and astonishing play of color. However, it requires special care to prevent abrasion or fracture.

Opal was highly prized by the ancients. Nonius, a Roman senator, chose exile rather than give up his opal to Mark Antony. However, in the nineteenth century, a foolish interpretation of a scene in a novel by Sir Walter Scott gave rise to the superstition that opals were unlucky. In recent years, their great beauty has redeemed them and they have won back much of their popularity.

In historic times, opal has been obtained from Czechoslovakia and, more recently, from Queretaro and Zimapan, Mexico, and Virgin Valley, Humboldt County, Nevada, which produces fine black opals. The most important source of present-day precious opal is from several areas in the interior of Australia. It is also found in Latch County and Bliss, Iowa; Horse Haven Hills, Washington; and Hart Mountain, Oregon.

OTHER GEMS AND DECORATIVE STONES

Feldspars

The feldspars are the most common minerals in the earth's crust, and yet only a relatively few occurrences yield clear, transparent material or colorful stones.

Orthoclase occasionally occurs as transparent, clear, or yellowish crystals or grains suitable for gems. The mean index is 1.52 to 1.58; the dispersion is also low, 0.012; hence, the stone has no notable fire, but it makes a pleasing faceted gem. Amazonite, the triclinic potash feldspar (found in Amelia County, Virginia; Florissant, Colorado; and Spruce Pine, North Carolina), sometimes has a pleasant green color; it makes good cabochon stones for rings, pins, and brooches. Moonstones (found in Media, Pennsylvania; Oliver, Virginia; and Black Range, New Mexico) are opalescent varieties of orthoclase (adularia) and oligoclase, a member of the plagioclase series of feldspars; they owe their opalescence to the presence of minute inclusions, which scatter the light. The most colorful of the plagioclase feldspars is labradorite (found in Essex and Lewis counties, New York, and Modoc County,

California), which occurs rarely in large grains in pegmatites. Its striking change of colors can be seen when the specimens are rotated in light from a lamp; it is caused by the reflection and refraction of light. Peristerite, a variety of albite, sometimes shows pleochromism.

Lapis Lazuli

Lapis lazuli is the name applied to a mixture of minerals. The material has been known since ancient times and has been used as a decorative inlay in mosaics, vases, and other ornamental objects. In modern times it is cut *en cabochon* for rings and brooches. It is also cut into spheres and used as decorative polished slabs. Lapis lazuli was formerly ground for the pigment ultramarine. The blue mineral of this gem material is lazurite, which has a hardness of 5 to 5.5 and a specific gravity of 2.4 to 2.5. The impure material is slightly heavier, because it contains calcite, haüynite, diopside, amphiboles, mica, and pyrite in varying amounts. Pyrite is the usual associate; it adds to the attractiveness and is an aid in identification, which can usually be made by sight alone. Lapis lazuli occurs in Afghanistan, Siberia, Chile, and in the United States in Gunnison County, Colorado, and Upland, California.

Turquoise

Turquoise—a hydrated phosphate of copper and aluminum—is one of the most attractive of the opaque to translucent gemstones. Its distinctive blue to greenish blue color provides a striking contrast with both silver and gold settings. Turquoise has been used as a gem and decorative material for thousands of years. It was highly favored by the American Indians for decorating their objects of worship, and today many tribes of the Southwest remain skillful at producing their typical handsome jewelry set with the stone.

Until recently the price of turquoise was relatively moderate, but the superb designs of American Indian jewelry have placed it in great demand, and the price of turquoise has increased correspondingly. The mineral is found near Nishapur, Iran; in the Sinai Peninsula; and in Australia and Turkestan. In the United States it occurs at many sites in the Southwest, where it is associated with extensive copper mines. Important localities are Fax Turquoise Mine, Cortez, and Royal Blue Mine, Nye County, Nevada; and King Mine, Conejos County, Colorado. (Other localities are noted in Chapter 7.)

Variscite

Variscite is a massive, fine-grained, hydrated aluminum phosphate containing some iron. It occurs in veins as fine-grained masses, flattened nodules, veinlets, and crusts. The hardness is 3.5 to 4.5, and the specific gravity, 2.57 to 2.61. Important occurrences are at Fairfield and Lucin, Utah; Emerald Claim, Esmeralda County, Nevada; Brazil; Germany; Austria; Czechoslovakia; and western Australia. The color—light to dark green and bluish green—is caused by the presence of chromium or vanadium. Variscite is cut *en cabochon* for pins and makes elegant cabinet specimens.

Rhodonite

Rhodonite is the triclinic metasilicate of manganese, with a rose red or pink color. The transparent to opaque mineral takes a good polish and is used as an ornamental stone as well as a cabochon for rings and brooches. The material is obtained from the Ural Mountains, Madagascar, Brazil, and Massachusetts; exceptional crystals come from Franklin, New Jersey. (Other localities are noted in Chapter 7.)

Rhodochrosite

Rhodochrosite is a carbonate of manganese. Most specimens are entirely too soft for use in jewelry, except for the fine, banded material from Catamarca Province, Argentina, which makes desirable cabochons for brooches. (Other localities are noted in Chapter 7.)

Jadeite and Nephrite

These two minerals, when in fine, interlocking acicular grains of a green color, constitute the mass of material known as jade. Jadeite is a sodium-aluminum silicate, with a hardness of 6.5 to 7 and specific gravity of 3.3. The color varies from white to greenish white and grass green. It is translucent to opaque, with an index of refraction of 1.66. Nephrite, a fine grained variety of actinolite, is a calcium-magnesium-iron silicate, with a hardness of 6

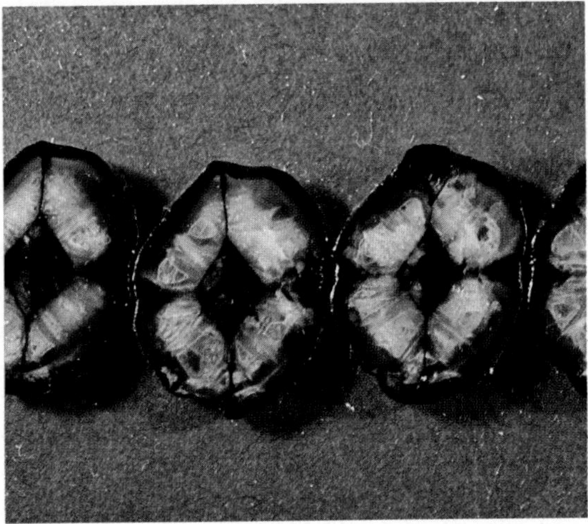

Top: Banded malachite (green) with chrysocolla (blue) from the southwestern United States. *Above:* Slices of chiastolite (a variety of andalusite) displaying a cross motif, from Lancaster, Massachusetts.

to 6.5 and specific gravity of 2.9 to 3.1. Its color varies from white to dark green. It is transparent to opaque and has an index of refraction of 1.62. Both jade minerals are the product of metamorphism and occur in association with other metamorphic minerals in schistose formations. They are tough and compact and take a good polish.

Jade is held in high esteem in the Orient, where it has traditionally been carved into elaborate designs. We know it was used by primitive peoples, as jade objects have been found at archeological sites in Switzerland, Egypt, Asia Minor, British Columbia, and South and Central America. During the past fifty years, many jade localities have been discovered along the Rocky Mountains, from Mexico to Alaska. Southeastern Asia has long been a source of the material. Jadeite can be found in Cloverdale and San Benito County, California; nephrite in Wausau, Wisconsin; the Kobuk region of Alaska; Curry County, Oregon; Monterey County, California; and south central Wyoming.

A few other green minerals resemble jade and have been given such names as "South African jade," "Vesuvianite jade," and "Serpentine jade"; however, all these names are erroneous and should not be used to confuse actual jade with other minerals.

Peridot

Peridot is an orthorhombic magnesium-iron silicate, the gem variety of the mineral olivine. Its hardness is 6.5 to 7, specific gravity 3.2 to 3.4, index of refraction about 1.63, and dispersion low, 0.018. Its color varies, from colorless to shades of yellow, brown, and green. The best gem color is yellowish green to green. The olive green variety is called peridot, and the greenish yellow, chrysolite. The gems look well either brilliant- or square-cut, although they are sometimes cut *en cabochon.* Gem peridot in the past came mostly from St. John's Island in the Red Sea. It is also found in Burma; Sri Lanka; and Brazil; Buell Park and Peridot Mesa, Arizona; and Afton, New Mexico.

Spinel

Spinel is a magnesium aluminate, with a hardness of 8, specific gravity of 3.5 to 3.7, index of refraction about 1.72, and colors ranging from red to blue, brown, black, and colorless. Iron, manganese, cobalt, and chromium impurities account for the variations in color. Many names have been given to this gem material, but the most accurate designations are on the basis of color, such as red spinel or green spinel. Spinels of gem quality come from the gravels of Thailand, Burma, and Sri Lanka. Other sources are Afghanistan, India, Madagascar, Australia, Brazil, northern New Jersey (Franklin, Newton, Sparta, and Hamburg), southern New York (Amity, Warwick, Edenville), and the Culsagee Mines, Franklin, North Carolina.

Chrysoberyl

Chrysoberyl provides two fine gemstones of striking properties. The mineral is orthorhombic and frequently twinned into pseudohexagonal combinations; it also occurs in loose, rounded grains. Chrysoberyl is a beryllium aluminate and may contain a little chromium. The hardness is 8.5, specific gravity 3.5 to 3.8, color yellowish green to green or sometimes yellow to brown. It is transparent to translucent and pleochroic in the green and brown varieties. When cut *en cabochon,* the silky, somewhat fibrous stones concentrate the reflected light in a line, which changes as the stone is turned. This is the true cat's-eye, and good specimens command high prices.

Another gem of this mineral is alexandrite, which is strongly pleochroic, showing grass green, columbine red, and orange yellow. The stone is grass green in daylight, columbine red in tungsten-lamp light.

Chrysoberyl originates in granite pegmatites and metamorphic gneisses and schists. Gemstones are found in Brazil; Sri Lanka; China; Saratoga Springs, New York; and Hedgehog Hill and Ragged Jack Mountain in Oxford County, Maine. Alexandrite is from Sri Lanka, the Ural Mountains, and Tasmania.

Zircon

Zircon is transparent to opaque, with colors ranging from brownish red to blue. The index of refraction is 1.92 and 1.98, the double refraction 0.06, the dispersion 0.048—properties that place zircon next to diamond in brilliance and fire. The gem name hyacinth is applied to the transparent yellow, orange, red, and brown varieties. Blue zircon, which is produced by heat treatment, is obtained from Indochina. Gem zircon occurs in Sri Lanka, New South Wales, and New Zealand. (Other localities are noted in Chapter 7.)

Apatite

The gem crystals of apatite are usually in hexagonal prisms. They are colorless, pink, yellow, blue, purple, or violet. They are transparent and vitreous, with an index of refraction of 1.64, a hardness of 5, and specific gravity of 3.1 to 3.2. The mineral is a calcium fluor- or chlor-phosphate. The brilliant- or square-cut gems are attractive, but are too soft for ring stones or brooches unless carefully handled. Gem material occurs in Germany; Switzerland; Sri Lanka; Mount Apatite, Harvard Quarry on Noyes Mountain, and Auburn, Maine; Pelham, Massachusetts; and Durango, Mexico.

Azurite and Malachite

Azurite and malachite are chiefly ore minerals of copper. They are too soft for use as jewelry; but as brightly colored minerals they are prized for their decorative effects and are popularly fashioned into such ornamental objects as tabletops, vases, and ashtrays, in addition to their use as cabinet specimens. The characteristic feature of malachite is its beautiful concentric banding in different shades of green. Both minerals are translucent to opaque and occur in the weathered parts of copper mines around the world. Some of the finest material came in former years from the Ural Mountains. At present it is mined in Katanga, Zaire; northern Rhodesia; and the southwestern United States. (Other localities noted in Chapter 7.)

Benitoite

This mineral crystallizes in the hexagonal system in good crystals; it has a hardness of 6.5 and specific gravity of 3.65. Its pale to deep blue color is caused by the presence of titanium, and it is strongly pleochroic; the best color effect is when the table of the cut stone is parallel to the vertical axis. The index of refraction is high, 1.757 and 1.804. It occurs in excellent crystals in San Benito, California.

Chrysocolla

This copper mineral has a lovely blue color that commends it as gem material, although it is fairly soft; it has a low index of refraction (1.50) and lacks even moderate cohesion unless highly silicified. The material is abundant in the upper levels of the open pit copper mines, such as the Inspiration Mine, Miami, Arizona. (Other localities are noted in Chapter 7.)

Sodalite

This mineral—a sodium-aluminum silicate—also has a fine blue color, but its hardness is only 5 to 6, specific gravity 2.2 to 2.4, and index of refraction very low, 1.48. In addition, its perfect cleavage causes it to break readily. It is nevertheless used as cabochons in brooches, since the effects of wear and breakage are offset by its lovely color and low cost. Localities include the Princess Mine, Bȧneroft, Ontario; Dennis Hill, Litchfield, Maine; Magnet Cove, Arkansas; Brazil; and Madagascar.

Spodumene

An ore mineral of lithium, spodumene has a medium hardness; it produces several fine gems with an index of refraction of 1.67 and low dispersion of 0.017. The clear to light yellow variety is known as spodumene; the rare yellowish green to grass green material is called hiddenite; and the pink to lilac variety, kunzite. Hiddenite is found at Stony Point, North Carolina; kunzite and spodumene gem minerals occur in pegmatites near Pala, California; Madagascar; and Minas Gerais, Brazil. (Other localities are noted in Chapter 7.)

Thomsonite

This orthorhombic mineral is a hydrated silicate of calcium, sodium, and aluminum. It has a hardness of 5 to 5.5, specific gravity of 2.3 to 2.4, and a low mean index of refraction, 1.51 to 1.54. It is white, reddish green, or brownish. It occurs as spherical concretions in basic igneous rocks as amygdules, and is cut chiefly *en cabochon.* It is found on the beaches of Isle Royale, Michigan; Thomsonite Beach, Cook County, Minnesota; the south shore of Lake Superior; in Scotland, Italy, and India.

Titanite

This monoclinic mineral often develops wedge-shaped and flattened crystals. It is a silicate of calcium and titanium, with a hardness of 5 to 5.5 and specific gravity of 3.4 to 3.6. It is yellow to green or brown, has an adamantine luster, and is transparent to opaque. The index of refraction is very high—1.90 to 2.03—the double refraction of 0.13 is strong, and the dispersion is very high, 0.050. With the exception of the low hardness, the other properties make titanite an outstanding faceted gemstone. The mineral occurs in metamorphic schists and limestones in Switzerland; also in other European localities; and in Maine; the Tilly Foster Mine, Brewster, New York; and Upper Chichester Township and John Mullen's Quarries, Chester, Pennsylvania.

Zoisite

This orthorhombic mineral has prismatic crystals and perfect cleavage; it is a basic silicate of calcium and aluminum. The hardness is 6

Above left: Cordierite, a strongly pleochroic gem mineral, from Richmond, New Hampshire. *Above right:* From the Moyen District in Chad, crystals of dioptase, a gem mineral found in copper deposits.

to 6.5, specific gravity 3.25 to 3.37, and mean index of refraction 1.703. It is yellowish brown, apple green, light red, and transparent to translucent. Zoisite has strong pleochroism in the pink variety—called thulite—which comes from Leksviken in Norway, Piedmont in Italy, and Tanzania. Excellent crystals with a large number of forms, colorless to blue to brownish, come from Tanzania and furnish the prized dark blue gem tanzanite. The blue color in some specimens is obtained by heat treatment. In the United States zoisite is found in the Putnam Mine, Mitchell County, North Carolina; Okanagan County, Washington; Douglas County, Nevada; and in the Black Hills of South Dakota.

Andalusite

This mineral occurs in good orthorhombic crystals, and the gem varieties are green, yellow green or light green, brown, pink, red, and violet. It has a vitreous luster and medium pleochroism; the mean index of refraction is 1.64. It is a silicate of aluminum. The faceted stones wear well when handled with care and are difficult to distinguish from tourmaline and brown apatite. The gems are usually step-cut. A twinned variety, chiastolite, reveals a dark central cross when cut *en cabochon* normal to the vertical axis. Andalusite occurs in the area near Lancaster, Massachusetts; in Delaware County, Pennsylvania; and in Madera, Mariposa, and Kern counties, California.

Cordierite (iolite, dichroite)

The orthorhombic mineral known as cordierite is a basic magnesium-iron-aluminum silicate, which occurs chiefly in metamorphic rocks. Its resistance to weathering separates it from the decomposed rock, and as a result it is deposited as pebbles in gravels. The strong pleochroism in dark blue, light blue, and yellowish colors makes it an attractive cabochon stone. It has a hardness of 7 to 7.5 and a specific gravity of 2.6. The mean index of refraction is low, 1.55, and double refraction is low, 0.008. Sri Lanka is the most important locality. The gem is rarely faceted and is a collector's item. It can be found in Hungry Horse Hill, North Haven County, Connecticut; Albany County, Wyoming; California; Idaho; and Colorado.

GEMS OF ORGANIC ORIGIN

A few gem materials, one highly prized, owe their existence to organic processes. The most valuable of these is the pearl, which originates within the mantle of mollusks. The early humans who used clams for food surely knew about the pearl and prized it as something unusual and valuable in trade. Pearls have been very highly prized by the rulers of the empires of the Middle East. Pearls develop when an irritant—such as a grain of sand—enters the mantle area of a mollusk. The animal at first attempts to reject the irritant, but being unsuccessful, it then secretes its shell-forming nacre about the object, continuing to build up the form until a pearl results. Another material of organic origin, also used from early times, is precious coral. A branching red carbonate structure is built up by the coral polyps in shallow parts of warm seas and recovered by divers.

Three other gem materials of organic origin are the result of plant activity. These are jet, a variety of hard lignite, anthracite coal, and amber. Jet occurs in the vicinity of Whitby, England, where it is carved into art objects and cabochons and beads for use as jewelry. In the United States it is found in Ann Arundel County, Maryland; El Paso County, Colorado; Santa Rosa, New Mexico; Presidio County, Texas; and Wayne County, Utah. The hard anthracite coal from the Scranton–Wilkes-Barre area of Pennsylvania (also found in Ashley, Summit Hill, and Nanticoke, and in Luzerne County) is used in the same manner. Amber is the most sought-after gem product of plants. It is formed from the consolidated, dried-out resin of coniferous trees that grew north of the Baltic Sea during interglacial periods. The trees are now extinct, but the amber formed from the sap exists in quantity within the glacial sands along the south shore of the Baltic. Good-quality amber enters the gem market for use especially in art objects and as beads. The non-gem material enters into various kinds of resin products. Amber is found in the United States in Nantucket and at Gay Head, Martha's Vineyard, Massachusetts; Staten Island, New York; Vincentown, New Jersey; Black Hills, South Dakota; and Ventura County, California.

Appendices

HABITS OF MINERAL GROWTH

Numerous terms are used to describe the habits of rocks and minerals. Unlike the precision of terms used in crystallography, these are general terms that describe approximate shape or texture, and there is occasionally some overlapping in the features they refer to. However, since certain minerals usually occur in a certain shape—or in several typical shapes—knowledge of a mineral's habits is valuable in determining its identity. For instance, rutile generally develops needlelike crystals, which may be described as *rutilated* or *acicular;* hematite may develop as *plates,* in *blades,* or in *reniform masses.*

Acicular. Acicular crystals are slender and needlelike. They may be randomly oriented within rocks, cavities, or on the walls of open fissure veins. An example is mesolite.

Amygdaloidal. The word means almond-shaped and refers to mineral-filled cavities in igneous rocks. The cavities are usually very irregular in shape and size. The most prevalent occurrence is in volcanic rock flows or near surface intrusions, which are frequently riddled with cavities carrying many different kinds of minerals.

Arborescent. Treelike or branching structures are usually designated by this term. Excellent examples are to be found among the fine native copper specimens from the Lake Superior copper area, on the Keweenaw Peninsula of northern Michigan. Another example is marcasite in clay from Folkestone, England.

Asteriated. Starlike reflections may be created by needlelike inclusions. An eye is produced by reflection from a single set of inclusions, and extends perpendicular to the needles. A star is created by sets of needles intersecting at some angle. Phlogopite often displays asterism.

Bladed. Bladed minerals develop elongated plates resembling knife blades or the leaves of certain lily plants.

Blocky. Crystals are said to be blocky when they are roughly equal in their three dimensions and have the appearance of small blocks. So formed, they are described as *equant,* having approximately equal dimensions. Many crystals in the cubic crystal system—such as galena and fluorite—can be described as blocky.

Botryoidal. The term means grapelike, referring to the many hemispherical aggregates clustered on the surface of a mineral specimen. This produces the appearance of a cluster of grapes. An example is limonite.

Capillary. This term refers to exceedingly slender, elongated hairlike crystals; they are thinner and usually more elongated proportionally than acicular crystals.

Cellular. Surface or near-surface magmas that contain large amounts of dissolved gases frequently develop open cavities, which are created by the expanding gases as the pressure is reduced, as when a bottle of carbonated beverage is opened. These rocks—such as basalt—are often made of glass and are described as cellular.

Clastic. Clastic specimens consist of fragments of preexisting rocks or minerals, which are clearly visible and may be cemented with one or more minerals—for instance, the quartz particles in sandstone and conglomerates, or the rock fragments in agglomerates.

Cockscomb structure. A few minerals develop a flat, almost semicircular arrangement of crystals, with the points radially arranged along the periphery, giving the effect of the edge of a cock's comb, as in marcasite.

Colloform. The term is derived

from "colloid," which describes material in a very fine state of division. Colloids frequently form globular masses when they agglomerate; these masses may later recrystallize into radial structures. However, it appears that most rounded forms of mineral aggregates are due essentially to growth of acicular crystals from a center. The crystals grow as the space availabe increases, giving the mass a spherical or, a hemispherical form.

Columnar. Crystals described as columnar resemble columns; they are usually rather thick and sometimes occur in parallel groups. Many columnar crystals are of a prismatic habit. An example is aragonite.

Compact. See *Massive.*

Concentric. This structure is made up of curved layers or bands of varying textures, colors, or mineral content. Malachite displays concentric banding.

Contorted. When deeply buried and highly heated, shales and similar rocks become soft and plastic. The small veins of minerals they contain may be twisted or crumpled in close folds; they are then described as contorted.

Coralloidal, flos ferri. Minerals with this character have a cylindrical shape in cross section and form twisted, forked rods that somewhat resemble a mass of coral. The best example is white aragonite from the iron mines of Styria, Austria, where it is called *Eisenblüte.* The unusual form of this mineral makes it a striking display piece.

Crustified. Minerals on the walls of fissures or other open spaces in rocks usually crystallize into a layer or crust, with terminations pointing into the open space. These structures are also referred to as *comb* structures.

Curved faces. Some crystals are distorted by an abnormal, complex internal grouping of their atoms plus an admixture of similar, extraneous atoms within the crystal lattice. This distorts the crystal planes and produces ill-shaped crystals with curved faces. Some minerals—such as dolomite and pyrite—tend to have curved crystal faces, a characteristic that helps to identify them.

Curvilaminar. This term is used to designate those platy minerals in which the laminae have been curved by the pressure of unequal forces. The minerals retain their flexed shape even after the pressures are equalized. An example is lepidolite.

Dendritic. This term describes the branching or fernlike structure assumed by some minerals. A good example is produced by pyrolusite, when it develops along the bedding planes of some sedimentary rocks—for example, in lithographic limestone, Solenhofen, W. Germany.

Divergent. Individual crystals in a radial structure are said to be divergent when they are clearly separated.

Drusy. The term applies to a lining of minute crystals on the walls of cavities in fissure veins. The crystals may all be of the same species, or they may be a mixture of species. Druses occur in openings in nearly all kinds of rocks and often yield fine crystals. Fine microcrystals frequently occur as druses in small pockets or cavities.

Earthy. Earthy materials resemble the usual appearance of ordinary soil. They are often the alteration products of earlier minerals or rocks and occur as a loose, crumbly mass. Some occurrences of bauxite and hydrated iron oxides have an earthy appearance.

Etched surfaces. Minerals with etched surfaces have been subjected to the solvent action of certain liquids or gases, or have failed in part to yield perfectly smooth surfaces at the end of crystallization. These imperfections on the crystal faces are small geometrical depressions, which follow a definite design for any particular crystal face. The symmetry of the figures is the same as that of the crystal. Etching appears on beryl.

Feathery, plumose. Resembling feathers, this phenomenon is exhibited sometimes by aggregates of mica grains that are in roughly parallel positions—as in the mineral muscovite.

Felty, leathery. This type of aggregation is probably the oddest in the entire field of mineralogy. Occasionally, the fibers of some minerals, chiefly amphiboles, that fill small rock fractures are so thoroughly intergrown, crisscrossed, and matted that the mass of flexible material closely resembles a piece of felt or soft leather.

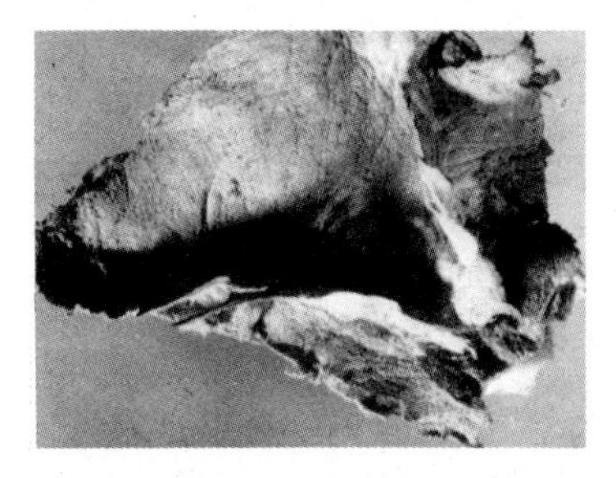

Fibrous. There are several minerals that can be cleaved into fine fibers. The most outstanding is asbestos or chrysotile, a variety of serpentine. The fine, silky fibers of this mineral are extremely flexible and can be spun into thread and woven into fireproof cloth.

Filiform. Derived from the word "filament," this refers to the threadlike or wirelike shapes of some minerals. It applies mostly to the native metals, some of which have roughly irregular, wirelike forms. Probably the commonest of these is wire silver.

Foliated. It is not uncommon for minerals to develop a platy or leaflike structure, called foliated, that can easily be split along the foliae. The micas, chlorites, and other similar minerals have such a structure.

Globular. Globular pieces are round or nearly round and vary in size. Some fine-grained sedimentary rocks contain rounded aggregates of minerals.

Irregular grouping. This is the common and usual occurrence of mineral groups. The arrangement is only a matter of chance and depends largely on the environment in which they develop.

Lamellar. This describes a mineral consisting of parallel plates or layers. These parallel plates appear on cleavage planes as fine parallel striations—in labradorite, for example. A synonymous term is *laminar.*

Leathery. See *Felty.*

Mammillary. Large, rounded masses resembling the mammae of mammals are referred to as mammillary. The masses are larger than those described as reniform.

Massive, compact. Solid, uniform masses without any visible crystal form or structure, such as clays and other exceedingly fine-grained materials, are referred to as massive or compact.

Micaceous (scaly). This term is frequently applied to schists or other rocks that are composed largely of very thin plates or scales of mica, usually muscovite or biotite.

Mirrorlike. Surfaces on some opaque crystals have faces that are so extremely smooth that they will reflect the images of nearby objects. Mirrorlike surfaces may be seen on pyrite.

Mossy. Minerals resulting from colloidal aggregation, usually siliceous, are sometimes subjected to penetrating growths of mosslike structures before they harden. These are produced by compounds of manganese and other elements.

Nodular. Minerals that have a globular habit, because of their compact, radiating grains or acicular crystals, may form small, rounded masses or nodules. Prehnite, marcasite, and other minerals are examples of this form.

Oölitic. An oölitic structure consists of small, rounded, concentric pieces about the size of fish roe, from which its name is derived. These small, rounded bodies are generated by wave action, which gathers coherent colloidal material sometimes about some minute body, such as a nucleus, by rolling it in water that contains colloidal particles.

Oscillatory growths. These growths, when poorly developed, appear as striations and are rather common on certain mineral species. Some of the most characteristic striations develop on some pyrite cube faces. These parallel ridges and furrows look like scratches, but are rather very narrow crystal faces of two crystal forms, the cube and the pyritohedron. The final growth of the crystal shifts from one of these forms to the other, thereby producing the ridge and valley character of the surface.

Parallel growths. A number of crystals develop so that their faces lie in parallel planes. This sort of growth is common on quartz crystals. One of the strikingly beautiful examples of parallel growth between two different species of minerals is that of tetragonal bisphenoids of chalcopyrite growing in parallel position on masses of sphalerite from the Kansas-Oklahoma lead-zinc area. In instances of this kind, the surface atoms of the base mineral exert some controlling force on the chalcopyrite atoms, causing them to assume precisely the same orientation at any spot on the sphalerite crystal.

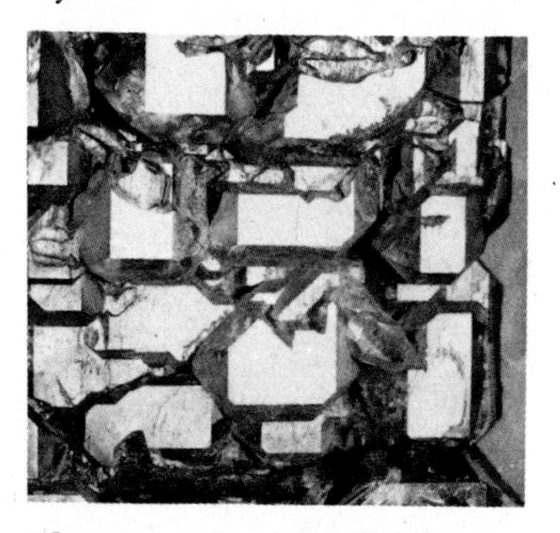

Phantoms. Occasionally the surfaces of a growing crystal are lightly coated with some extraneous material that provides a visible outline of the crystal at some period of its growth. The process is often repeated in a series, producing a number of forms or phantoms that are visible in a transparent crystal, such as quartz.

Pisolites. The structure of pisolites is the same as that of oölites, but the globular pieces are much larger, about the size of peas. In distinguishing the two structures one must consider the size, for pieces of intermediate dimensions also occur. It is best to give a numerical size for the individual pieces. Pisolites are found in bauxite.
Platy. Some minerals with large basal pinacoids and very scant prisms or pyramids occur as thin plates; in some instances they are paper-thin and extremely delicate. Hematite crystals are sometimes platy.
Plumose. See *Feathery*.
Prismatic. This is a term that applies to crystals that have developed prominent prism faces. However, a given mineral may not always have a prismatic habit, as it may grow in different ways. Such changes are brought about by variations in the physical-chemical nature of the solutions. Prismatic crystals vary in length. They occur frequently in quartz.

Pyramidal. Some crystals, such as zircon and rutile, habitually develop pyramid faces with narrow prisms. Octahedrons are often mistaken for pyramidal forms.
Radial. A radial structure, usually seen in globular or stalactitic masses, displays radiating lines when the mass is broken across. The term also applies to groups of crystals closely arranged in a radial pattern. When the pattern is open, like the spokes of a wheel, the arrangement is sometimes referred to as divergent. Goethite may grow in radial groups.

Reniform. The term describes a resemblance to a kidney; however, it includes many similar rounded shapes, up to the size of several inches, until they are large enough to be described as mammillary. An example is hematite.

Reticulated. The name refers to a latticelike or network structure, in which elongated or bladed crystals cross repeatedly in a regular pattern. Cerussite and rutile are among the best reticulated minerals.
Rosette. One of the most attractive mineral structures is the rosette, so named because it resembles a rose. It is built up by thin plates of a mineral in overlapping positions, which are distributed in a circular pattern. The most nearly perfect and prized of these is the *Eisenrose* of Switzerland, which consists of brilliant overlapping plates of hematite.

Rough. Normally, crystal faces are smooth when they develop free from interference with other growing crystals. However, frequently there is serious competition for space, in which case the faces of a crystal are subject to many interferences from other minerals. This causes them to become uneven, overgrown, pitted, or rough—as in pyrite.
Rutilated. Hairlike crystals of rutile may develop in radiating groups or irregular masses. When they are subsequently incorporated in the body of growing crystals of quartz, the resulting specimens are known as rutilated quartz.
Scalenohedral. The scalenohedron is a crystal form in the hexagonal crystal system. This form is often the prominent termination on calcite. Such calcite crystals are often exceptionally fine and are prized by collectors. Other crystal form names are also used to describe minerals, such as octahedral, rhombohedral, or sphenoidal.

Scaly. See *Micaceous*.
Scepter. The phenomenon is found chiefly in connection with quartz crystals, in which the early predominant prism form has almost ceased to develop. This allows the terminal positive and negative rhombohedrons to expand to a diameter greater than that of the original prism, giving the illusion of a prism supporting a larger, doubly terminated quartz crystal.

Schistose. This is a term applied to rocks in which the principal minerals are one or several micas arranged in roughly parallel positions. The easy and eminent cleavage of the mica produces a rock that splits easily along the plane of the mica.

Sheaflike. This odd mineral structure is best developed in the mineral stilbite, which resembles a sheaf of grain bound in the middle.

Silky, satiny. A few minerals that are built up of many fine, thin, elongated grains have a silky or satiny appearance, such as the satin spar variety of gypsum, or the gem material tiger's-eye.

Skeletal. Skeletonized crystals, resembling the bone structure of an animal, are a common occurrence for some minerals. The form has many variations.

Stellate. Crystals may form a radial grouping in which the individuals are distinctly separated. Groups of crystals arranged in this manner are known as stellate patterns.

Striated. Crystals are sometimes marked by straight, parallel lines or ridges, which are produced by the intersection of two crystal forms. An example is striated pyrite crystals, on which the intersection of the cube and pyritohedron faces makes the parallel striations, which conform to the symmetry of the crystal.

Stubby. Stubby crystals are short, prismatic, and pyramidal crystals or have other nearly equant dimensions. Some crystals of the isometric, tetragonal, and hexagonal crystal systems are inclined to be stubby. Quartz is a typical example.

Tabular. Some minerals develop thin crystals with two relatively large parallel faces. Such crystals are said to have a tabular habit. An example is barite.

Terminated. Crystals growing in an open space may be bounded by plane surfaces called faces. When these cover a substantial portion of a crystal—as in quartz—it is said to be terminated (by faces).

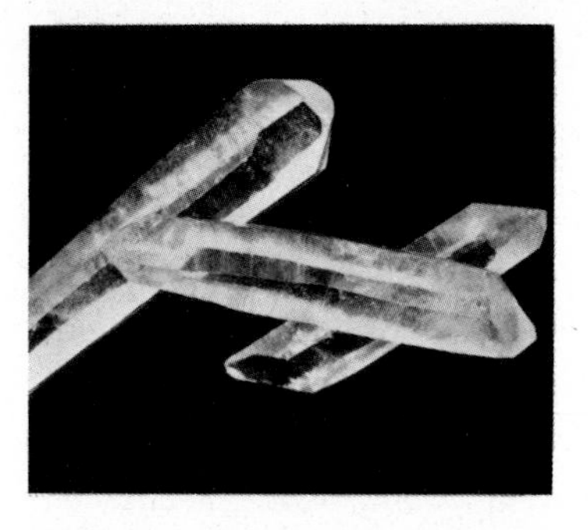

Vesicular. The term refers to the bubblelike cavities sometimes found in surface flows of igneous rocks. Viscous flow causes the bubbles to be elongated or drawn out into thin tubes.

Vicinal forms. Growth surfaces may appear on crystals in positions that do not correspond to the positions of prominent crystal faces. These surfaces are called vicinal, as in fluorite.

Vuggy. A vug is a small, open cavity in rock of any kind. The walls of the cavity are often lined with various minerals, which usually provide excellent crystal specimens for the micromounter, as well as very fine larger-sized specimens.

Warty. A warty specimen consists of small, globular groups of minerals having a distinctly rough surface instead of the customary smooth one. Psilomelare may develop warty surfaces.

Zoned. The structure, color, or distribution of inclusions may sometimes be concentrated in layers through a crystal. Such distributions produce a zoned appearance. Amethyst crystals often show a striking zonal arrangement of the purple color.

TESTS FOR THE CHEMICAL ELEMENTS

Blowpipe and chemical reactions are chiefly qualitative in nature; however, in some minerals a general knowledge of the proportional part of a few elements can be estimated roughly. There are two kinds of tests: dry tests made with the mineral alone in conjunction with known dry chemicals; and treating the powdered mineral and its elemental constituents with liquid reagents. Both kinds of tests are carried out with various procedures.

The more important tests are described below in detail sufficient to permit a qualitative recognition of the common elements. In actuality, many interferences from the associated elements will be encountered. It will often be necessary to effect some separations by precipitation and filtering to reach a test for the various constituents. Much time, patience, ingenuity, and skillful manipulation are necessary to reach the desired proficiency with the blowpipe and semi-microchemical methods of studying the chemical nature of minerals. But once the art is mastered, the use of the blowpipe, along with the physical properties of the minerals, will invariably lead to the identification of a large number of the common minerals.

The following information will be useful before the tests are attempted:

The spectroscope is a device that breaks up white light into its component colors. When a substance is ignited in a flame, the elements it contains give off intense colors; these are visible as bright colored lines superimposed against a continuous spectrum in the spectroscope. A given element may produce several lines, whose positions in the spectrum are characteristic and diagnostic for that element.

At a temperature of 4°C—the point at which water reaches its maximum density—one cc of water weighs exactly one gram and occupies a volume of one ml.

To determine whether a solution is acid or alkaline, insert a piece of litmus paper. If the solution is acid, the paper will turn red; if alkaline, the paper will turn blue.

The alkali metals are cesium, lithium, sodium, potassium, and rubidium. The alkaline earth metals are barium, calcium, magnesium, and strontium.

The rare-earth elements constitute a group with very similar chemical properties and they are extremely difficult to separate. Tests for these elements are usually not possible in the home laboratory.

The noble gases—helium, neon, argon, krypton, xenon, and radon—are chemically inert and do not tend to form compounds. They are therefore not to be expected in minerals, and tests for their presence are superfluous.

The following abbreviation is used:

HCl hydrochloric acid

Aluminum (Al)

This element is the third most abundant in the earth's crust and is an important constituent of a large number of minerals. In many minerals, aluminum is tightly bound chemically, and thus usually requires the fusion of the minerals to produce soluble compounds in order to obtain satisfactory tests.

Ignition with cobalt nitrate. Light-colored, infusible aluminum minerals assume a blue color when moistened with cobalt nitrate and intensely ignited. The test can be applied only to the infusible aluminum minerals, as the cobalt oxide obtained by ignition is often dissolved in the glasses of fusible minerals and thereby colors them a distinct blue. An excess of cobalt nitrate may readily obscure the blue color by a dense coating of black cobalt oxide.

When zinc silicates are moistened with cobalt nitrate and strongly ignited, they may yield a blue color, but when heated with sodium carbonate on charcoal they produce a white coating.

Precipitation with ammonia. A white gelatinous precipitate of aluminum hydroxide, $Al(OH)_3$, is produced when an acid solution containing aluminum is made alkaline by the addition of ammonium hydroxide, NH_4OH. This precipitate can be distinguished from the red hydroxide of iron, $Fe(OH)_3$, formed in the same way, as the former is soluble in a warm solution of potassium hydroxide, KOH, while the latter is insoluble in KOH.

Ammonium (NH_4)

A few minerals contain the ammonium radical, which plays the part of a metal in its compounds. Compounds containing NH_4 yield ammonia, NH_3, which can be detected by its characteristic color when boiled with a solution of potassium hydroxide in a test tube, or when heated in an open tube with sodium carbonate, Na_2CO_3, or calcium carbonate, $CaCO_3$. When ammonia fumes and those of HCl are brought into contact, dense white fumes of ammonium chloride, NH_4Cl, result.

Antimony (Sb)

When antimony and its compounds are heated before the blowpipe in the reducing flame, antimony volatilizes and gives a pale greenish color to the flame.

Oxidation on charcoal. When most antimony minerals are heated in the oxidizing flame on charcoal, they yield a dense white sublimate of antimony trioxide, Sb_2O_3, which is formed near the assay. The sublimate is very volatile and can be made to change its position on the charcoal. Where the coating is thin, it has a bluish color. The fumes lack a characteristic odor (differing from arsenic).

Test in open tube. When stibnite and other antimony sulfides are roasted in the open tube, a sublimate of oxides of antimony collects on the walls of the tube. In the presence of sulfur, the oxide settles along the bottom of the tube for some distance. A short distance above the mineral a powdery ring forms on the wall of the tube. The ring near the mineral is a volatile coating of antimony trioxide, which becomes yellow when hot and white again when cold. It is usually the most conspicuous.

Sublimate on plaster tablet. When mixed with a one-to-one mixture of potassium iodide and sulfur and heated on a plaster tablet, powdered antimony minerals give an orange red sublimate. The color disappears when exposed to strong ammonia fumes.

Oxidation with concentrated nitric acid. Antimony and its sulfides are oxidized by nitric acid, NHO_3, to metantimonic acid, a white precipitate very insoluble in both water and nitric acid. This property, useful in effecting a separation of antimony from its associated elements, may be tested for in the filtrate.

Arsenic (As)

When arsenic is volatilized before the blowpipe in the reducing flame, it imparts a violet tinge to the flame.

Roasting on charcoal. Arsenic, sulfides of arsenic, and arsenides when roasted on charcoal produce an exceedingly volatile white coating of arsenic trioxide, As_2O_3, which forms on the charcoal at some distance from the point of oxidation. The fumes given off in the reducing flame have the characteristic garliclike odor.

Sublimate on plaster tablet. When arsenic minerals are heated with a one-to-one mixture of potassium iodide and sulfur, a very volatile sublimate occurs as a yellow to orange coating of arsenic iodide on the tablet.

Oxidation in the open tube. Arsenic, arsenides, and sulfarsenides when heated slowly in the open tube yield a very volatile white sublimate of As_2O_3, that may show well-defined octahedral microcrystals. If the heating is too rapid, a metallic arsenic mirror may result instead of the arsenic oxide.

Heating in the closed tube. Arsenic and many arsenides produce a bright metallic sublimate known as an arsenic mirror. If the bottom of the tube is opened below the mirror and heated, the metallic arsenic becomes oxidized and gives the characteristic garliclike odor.

Reduction of arsenates and oxides in the closed tube. Powdered arsenates and oxides, as coatings on charcoal, may be mixed with powdered charcoal and introduced into the closed tube and heated. The hot charcoal reduces the arsenates and oxides, and the metallic arsenic is volatilized, forming a characteristic arsenic mirror near the point of reduction.

Barium (Ba)

Flame coloration. Most barium minerals, except silicates, impart a yellowish green coloration to the blowpipe flame. Moistening the test piece in hydrochloric acid, HCl, sometimes intensifies the color. The flame colors of boron and phosphorus should not be mistaken for that of barium. If strontium is present, the test piece after being moistened in HCl will show the red flame before the yellowish green color, because of the barium.

Alkaline reaction. As barium is one of the alkali earth metals, minerals of this metal, except the phosphates and silicates, will, when intensely ignited before the blowpipe, yield an alkaline reaction on moistened tumeric paper.

Precipitation as barium sulfate. Barium minerals soluble in hydrochloric or nitric acids will yield, from dilute solutions of these acids, a white precipitate upon the addition of a few drops of dilute sulfuric acid or a soluble sulfate.

Insoluble silicates must first be fused with sodium carbonate and the fusion then dissolved with hydrochloric acid. After dilution with half again as much water, the addition of a few drops of dilute sulfuric acid will precipitate the barium present as barium sulfate.

Bismuth (Bi)

Reduction on charcoal. When mixed with sodium carbonate, Na_2CO_3, and heated on charcoal, finely ground bismuth minerals produce a lemon yellow coating surrounded with a white border. A reddish white, easily fusible metallic button is obtained. It is lead gray when hot but becomes covered with an oxide when cold. Although the button is relatively brittle, it can be flattened somewhat by hammering but soon breaks up into many small pieces.

Iodide flux on charcoal. When the mineral powder is mixed with iodide flux and heated upon charcoal, a yellowish sublimate with crimson border is produced.

Sublimate on plaster tablet. When mixed with iodide flux on a plaster tablet, the finely powdered mineral produces a chocolate brown sublimate, which changes to a bright red when exposed to strong ammonia fumes.

Cadmium (Cd)

Reaction on charcoal. Cadmium minerals when mixed with sodium carbonate yield before the blowpipe on charcoal a sublimate, which is reddish brown near the assay and yellowish green at a distance. When the sublimate is very thin it may show an iridescent tarnish. In the presence of zinc, a common associate, the cadmium is volatilized, yielding a sublimate before that of zinc appears.

Calcium (Ca)

Alkaline reaction. When ignited before the blowpipe, calcium minerals become alkaline, and when placed on tumeric paper they turn brown. Exceptions are the silicates, phosphates, borates, and the salts of a few rare acids. Other minerals of the alkalies and alkaline earths produce a similar reaction; hence, their absence must be established for the test to be conclusive.

Flame test. Many calcium minerals when powdered, moistened with HCl, and introduced into the nonluminous flame of the Bunsen burner will yield a yellowish red flame. The strontium flame is redder and more persistent. The calcium flame appears as a flash of greenish yellow through the first division of the Merwin color screen; lithium and strontium differ in this respect.

Precipitation as calcium sulfate. The addition of a few drops of sulfuric acid, H_2SO_4, to a concentrated HCl solution of the mineral will precipitate calcium sulfate, $CaSO_4$. Dilution of the solution with water and the application of heat dissolve the precipitate (distinct from barium and strontium). Calcium sulfate will not form in a dilute aqueous solution.

Precipitation as calcium oxalate. When ammonium oxalate, $(NH_4)_2C_2O_4$, is added to an alkaline or slightly acid solution, a finely divided precipitate of calcium oxalate, CaC_2O_4, is obtained. The precipitate is coarser when precipitated from a hot solution. When allowed to stand for at least an hour it may be readily filtered.

Precipitation as calcium carbonate. A calcium solution made strongly alkaline with ammonium hydroxide, NH_4OH, will precipitate calcium carbonate, $CaCO_3$, upon the addition of ammonium carbonate. The precipitation is practically complete when made from a boiling solution; the precipitate can be removed by filtering.

Carbon (C)

Effervescence with acids. Carbonates are readily soluble in dilute acids with effervescence; however, in some instances the mineral must be finely powdered or the acid heated but kept well below the boiling point to avoid obscuring the bubbling caused by the liberation of CO_2, carbon dioxide. CO_2 may be determined by its inability to support combustion, by dropping a lighted splinter into the test tube where the carbonate is decomposed by an acid, or the liberated gas can be directed into an aqueous solution of calcium hydroxide, $Ca(OH)_2$, which becomes turbid by forming a calcium carbonate precipitate with the CO_2.

Heating in a closed tube. When heated in a closed tube, hydrocarbons such as asphaltum, bituminous coal, gilsonite, and similar materials undergo destructive distillation and emit tarry compounds, which condense on the sides of the tube. The remaining residue, if any, consists largely of carbon and any inorganic substances not decomposed by heat.

When carbonates are placed in the closed tube and intensely ignited, they are decomposed with the liberation of carbon dioxide, CO_2, which can be identified by directing the gas through a solution of barium hydroxide from which barium carbonate, $BaCO_3$, will be precipitated.

Chlorine (Cl)

Precipitation with silver nitrate. The addition of silver nitrate, $AgNO_3$, to a water or nitric acid solution of a chloride yields a white, curdy precipitate of silver chloride, AgCl. The test is very delicate; traces of chlorine produce a milky solution, which on exposure to light soon acquires a violet color. Silver chloride is soluble in ammonium hydroxide, NH_4OH. Bromine and iodine give similar reactions. Insoluble minerals are fused with sodium carbonate and dissolved in nitric acid.

Evolution of chlorine. A mixture of the powdered mineral with about four times its volume of potassium bisulfate and a pinch of powdered pyrolusite, MnO_2, is heated in a small test tube. The chlorine gas evolved may be recognized by its pungent odor or its bleaching action on a strip of moistened litmus paper held inside the tube. Insoluble chlorides must first be fused with sodium carbonate, Na_2CO_3, and dissolved in nitric acid.

Flame coloration with copper oxide. A salt of phosphorus bead is first saturated with copper oxide, CuO, and then brought in contact with the powdered mineral. In the presence of a chloride in the unknown, the bead when held in the nonluminous flame of the Bunsen burner will tinge the flame with an azure blue color due to the copper chloride, $CuCl_2$, that will be formed. Bromine gives a similar reaction.

Chromium (Cr)

Test with borax bead. The addition of a small amount of a chromium mineral to a borax bead will produce a decided yellow color in the hot bead in the oxidizing flame, the color changing to yellowish green when cold. In the reducing flame, the bead assumes a fine green color, which remains when cold.

Test with salt of phosphorus bead. The addition of a small to medium amount of the mineral to a salt of phosphorus bead produces a bead with a dirty green color while hot in both the oxidizing and reducing flames; when cold in both instances, the color of the bead is a fine green.

Precipitation with lead acetate. If insoluble in acids, the mineral is fused with sodium carbonate, Na_2CO_3, and potassium nitrate, KNO_3, in a platinum spoon or on charcoal. The fusion is dissolved in water made slightly acid with acetic acid. To the solution a few drops of lead acetate, $Pb(C_2H_3O_2)_2 \cdot 3H_2O$, are added, forming the yellow precipitate, lead chromate.

Oxidation with hydrogen peroxide. The fusion made in the above reaction is dissolved in water and acidified with dilute sulfuric acid, H_2SO_4. Hydrogen peroxide added to the cold solution produces a fleeting blue color due to the formation of the very unstable perchromic acid, H_3CrO_8.

Cobalt (Co)

Test with borax and salt of phosphorus beads. When cobalt minerals are fused in the borax and salt of phosphorus beads, they produce a fine dark blue color in both beads in both the oxidizing and reducing flames, when either hot or cold.

The interference of copper and nickel may be eliminated by fusing a borax bead of the mineral on charcoal with a small piece of metallic tin in the reducing flame. When the nickel and copper are reduced to the metallic stage, the blue color of the cobalt appears.

Columbium (niobium) (Cb)

Salt of phosphorus bead. The salt of phosphorus bead is pale yellow in the oxidizing flame while hot, colorless when cold; it is blue violet or brown in the reducing flame, depending upon the amount present. The addition of ferrous sulfate, $FeSO_4$, changes the color to red.

Reduction with tin. Finely powdered columbates are readily decomposed when heated to a dull red heat with potassium bisulfate, $KHSO_4$. The decomposition can be made in a crucible. The procedure is to mix the powder with about ten times its volume of the reagent and to place the mixture in a Pyrex test tube, fusing it over a Bunsen burner. After the fusion is complete, the melt should be distributed on the sides of the tube by slowly rotating it as the melt solidifies. In this form the melt is dissolved easily by the addition of hydrochloric acid, forming a chloride solution of columbium. When metallic tin is added and the solution is boiled, a light blue solution due to columbium appears. The addition of water heightens the color.

If zinc is used as the reducing agent rather than tin, the color may appear momentarily blue but change quickly to brown.

Copper (Cu)

Flame test. Copper sulfides should be roasted in the oxidizing flame before making the flame tests. Oxides of copper color the nonluminous flame of the Bunsen burner a vivid pure green; when the specimen is moistened with HCl, the flame is colored an intense azure blue tinged with green, and when moistened with hydriodic acid the flame is a fine green color.

Test with borax bead. A borax bead containing copper oxide heated in the oxidizing flame is green when hot and blue when cold; when heated in the reducing flame, the bead is colorless to green when hot and turns opaque red when cold.

Test with salt of phosphorus bead. In the salt of phosphorus bead, the color in the oxidizing flame is green when hot and a pale blue when cold. In the reducing flame, the bead is brownish green when hot and opaque red when cold.

Reduction to metal on charcoal. When fused on charcoal in the reducing flame, in a flux consisting of equal parts of sodium carbonate and borax, copper minerals readily produce globules of copper. Copper minerals containing arsenic, sulfur, or antimony should be roasted first to remove these substances. The metals reducible with difficulty remain in the slag, while the accompanying, easily reducible metals alloy with copper in the globule. Such alloys may be suspected by the color of the metal and may be tested for the individual metals.

Blue solution with ammonium hydroxide. Acid solutions of copper, when made alkaline with ammonium hydroxide, turn a deep azure blue. Although alkaline solutions of nickel are light blue, the color is concealed by the presence of copper.

Gold (Au)

Detected by physical properties. Metallic gold can be recognized so readily that blowpipe or chemical tests are almost unnecessary. The nuggets and very small pieces of the metal found in placers are identified easily in the gold pan by their high specific gravity and bright yellow color. In order to identify gold included in quartz and in other minerals, the substance is first finely ground in a steel mortar and the resulting powder is washed in the gold pan. Small, heavy, bright yellow specks readily identify the gold. The identity can be further established by hammering the small pieces into thinner plates, as the gold is exceedingly malleable.

On plaster per se. Compounds of gold are limited to the tellurides. When these are heated intensely before the blowpipe on either charcoal or plaster or in the forceps, they are decomposed, thus freeing the gold, which appears as bright yellow globules that are exceedingly malleable. Tellurium sublimate appears on the plaster and charcoal about the test specimen.

A small piece of gold when heated intensely for several minutes on plaster yields a slight purplish rose-colored coating near the assay. The color appears strongest on the cold tablet.

Cassins purple test. Gold can be dissolved readily in aqua regia and obtained as an aqueous chloride solution by evaporating the solution of gold to dryness and taking up the residue with water. To the aqueous solution, a few drops of freshly prepared stannous chloride are added. This will produce a finely divided precipitate, which appears purplish in transmitted light and brownish by reflected light.

Iodine (I)

Closed tube reaction. When mixed with potassium bisulfate, $KHSO_4$, in a closed tube, the powdered mineral liberates violet vapors of iodine, which often are accompanied by a sublimate of metallic iodine.

Precipitation with silver nitrate. A dilute nitric acid solution of an iodide to which have been added a few drops of silver nitrate yields a precipitate of silver iodide, AgI, which is nearly insoluble in ammonium hydroxide (as distinct from chloride and bromide).

Iron (Fe)

Tests with magnet. Although there are many iron-bearing minerals, only a few are naturally magnetic—magnetite, pyrrhotite, and pentlandite. Many become magnetic upon cooling after decomposition in the reducing flame. Cobalt and nickel minerals may produce magnetic assays when the reduction has proceeded far enough for some metal to accumulate.

Borax bead test. In the oxidizing flame, the borax bead is yellow to orange red when hot and yellow upon cooling; in the reducing flame, the bead is green when hot and light green when cold.

Precipitation with ammonium hydroxide. When ammonium hydroxide is added to a solution containing ferric iron, the iron is completely precipitated as a brownish red flocculent ferric hydroxide. The precipitate can be filtered readily and the iron completely removed from solution. Ferrous iron solutions can be oxidized easily to a ferric condition by boiling the solution with a small amount of nitric acid.

Lead (Pb)

Reduction on charcoal. A lead globule may be obtained from lead minerals by mixing the powdered mineral with an equal part of charcoal and three parts of sodium carbonate. The assay is moistened with water to make a paste, which is transferred to charcoal and heated in the reducing flame.

The small lead particles may be collected into a single globule that appears bright in the reducing flame, but turns dull when the flame is removed and the button cooled. As lead is somewhat volatile, a sublimate of lead oxide accumulates on the charcoal about the assay. The color is sulfur yellow near the assay and bluish white at a distance. The coating is volatile and can be moved about with the blowpipe flame. The lead button is soft, malleable, and sectile.

Lead minerals that contain antimony should be roasted in the oxidizing flame before reduction with sodium carbonate and charcoal.

Precipitation with sulfuric and hydrochloric acids. Lead minerals go into solution most readily in dilute nitric acid. From the nitrate solutions lead is precipitated as a white insoluble compound, lead sulfate, $PbSO_4$, by the addition of a few drops of sulfuric acid. When a few drops of hydrochloric acid are added to a similar amount of cold nitric acid solution, a white precipitate, lead chloride, $PbCl_2$, is formed, which dissolves when heated.

Sublimate on charcoal. When the powder of lead minerals is mixed with potassium iodide and sulfur and heated on charcoal, a greenish yellow sublimate is produced on the charcoal.

Lithium (Li)

Flame test. Lithium imparts a clear crimson color to the nonluminous flame of the Bunsen burner and blowpipe. The color is similar to but not as persistent as that of strontium, for lithium is more readily volatilized from its compounds than strontium. As the volatility of barium is intermediate, in a lithium solution containing barium chloride, $BaCl_2$, the red color of lithium appears before the green of barium, and when barium is present in strontium solutions, the green color of barium appears before the red of strontium.

Manganese (Mn)

Borax bead test. The borax bead containing manganese becomes reddish violet in the oxidizing flame. In the reducing flame it becomes colorless. The salt of phosphorus bead gives similar but less sensitive reactions.

Sodium carbonate bead test. When an oxide of manganese is dissolved in a sodium carbonate bead and heated in the oxidizing flame, the bead assumes a green color when hot and a bluish green color upon cooling. The cold bead is not transparent. In the reducing flame the bead becomes colorless. The test is a very delicate one because other substances are not likely to interfere.

Oxidation to permanganic acid. In a manganese solution boiled with concentrated nitric acid and lead oxide, PbO_2, the manganese is oxidized to a purplish permanganic acid, which appears in the supernatant solution after the lead oxide has been allowed to settle or has been removed by filtration. The filtrate will appear to have a purplish hue.

Heating in closed tube. Some of the higher oxides of manganese yield oxygen when heated intensely in the closed tube. The liberated oxygen may be detected by introducing a glowing splinter into the tube, which will burn briskly in the presence of liberated oxygen.

Mercury (Hg)

Sublimate on plaster tablet. When mercury minerals are finely powdered and mixed with sodium bisulfite, potassium iodide, and sulfur, they will yield a multicolored sublimate when heated on a plaster tablet. The colors are a combination of yellow, crimson, and greenish black.

Molybdenum (Mo)

The minerals of molybdenum are limited to a relatively few species, the chief of which is molybdenite, MoS_2. Among the oxidized products are wulfenite, $PbMoO_4$. Both these minerals have characteristic appearances and are readily identified. In testing the minerals for the presence of molybdenum, a different treatment is required.

Flame test. A fragment of molybdenite held in the forceps and heated in the reducing flame imparts a pale yellowish green color to the flame.

Heated on the plaster tablet. Per se the mineral produces a white sublimate of molybdic oxide. When this is touched with the reducing flame, the white oxide is turned deep blue. When a molybdenum mineral is heated with a mixture of potassium iodide and sulfur, a deep blue sublimate forms.

Bead test. Molybdenum oxide added to the salt of phosphorus bead yields a green-colored bead in the reducing flame.

Heating in the open tube. The powdered mineral when heated in the open tube affords a network of white crystals of MoO_3 near the assay. If the crystal-bearing section of the tube is removed and broken to expose the crystals to the reducing flame of the blowpipe, they change to a blue color.

Heating with sulfuric acid. If some finely powdered molybdate is added to a few drops of concentrated sulfuric acid in a casserole, heated until it fumes strongly, and allowed to cool, the mass will acquire an intense blue color.

Sublimate on charcoal. When thin fragments of molybdenite are heated on the flat surface of a charcoal block for one minute in the oxidizing flame, a coating of molybdic oxide, MoO_3, is produced a short distance from the material. The sublimate is pale yellow when hot, white when cold, and an intense blue when touched with the reducing flame. Nearer the assay, a thin copper-colored coating appears, which is best observed in reflected light.

Nickel (Ni)

Bead test. In the oxidizing flame, a borax bead containing a nickel salt yields a violet color when hot and a reddish brown when cold. In a salt of phosphorus bead, the color is yellow but the test is not very satisfactory. Cobalt, unless in very small amounts, obscures the nickel color.

Test with ammonium hydroxide. When a somewhat concentrated acid solution of nickel is made basic by the addition of ammonium hydroxide, a slight precipitate of nickel hydroxide, $Ni(OH)_2$, is formed and is quickly dissolved, imparting a pale green color to the solution. Copper gives a similar but far more intense color.

Precipitation with dimethylgloxime. The powdered mineral is dissolved in nitric acid and the solution made alkaline with ammonium hydroxide. After filtering, if necessary, a small amount of an alcoholic solution of dimethylgloxime is added to the ammoniacal solution; it will produce a pronounced scarlet crystalline precipitate of nickel dimethylgloxime.

Phosphorus (P)

Flame test. A number of phosphates impart a bluish green color to the blowpipe flame; others will produce the same reaction when moistened with sulfuric acid and then heated.

Reaction with ammonium molybdate. The mineral is dissolved in nitric acid, after being fused with sodium carbonate if necessary, and a few drops of the solution are added to a solution of ammonium molybdate. If phosphorus is present, a yellow precipitate of ammonium phosphomolybdate forms in a few minutes. The addition of a little ammonium nitrate to the solution hastens the formation of the precipitate.

Potassium (K)

Flame color. Many potassium minerals impart a clear, pale violet color to the flame. However, the nonvolatile silicates must first be powdered and m xed with an equal amount of powdered gypsum. The mixture must be moistened with hydrochloric acid and introduced into the flame on the platinum wire. Sodium is usually present with potassium, and its strong yellow flame is likely to obscure the potassium flame. The yellow sodium flame can be filtered out by using a piece of blue (cobalt) glass or a Merwin color screen. Cesium yields a flame of the same color, but it can be distinguished easily with a direct vision spectroscope.

Alkaline reaction. After intense ignition, some potassium minerals give an alkaline reaction when placed on moistened tumeric paper.

Precipitation with platinic chloride. The addition of platinic chloride, $PtCl_4$, to a neutral or slightly acid solution of a potassium salt results in the formation of a yellow precipitate of potassium platinic chloride, K_2PtCl_6.

Silicon (Si)

Formation of a silica gel. After the mineral is dissolved in nitric acid (fused with sodium carbonate if necessary) and evaporated almost to dryness, a jellylike mass of silica will appear if the mineral is a silicate. In the test tube it will be noted clinging to the sides of the tube as the evaporation approaches completion.

Silver (Ag)

Precipitation with hydrochloric acid or a chloride. The mineral is powdered and taken into solution with nitric acid (fused with sodium carbonate if necessary) and filtered if a white precipitate has formed during the solution with nitric acid. To the filtrate a few drops of hydrochloric acid or a chloride solution—such as sodium chloride or ammonium chloride—are added. If silver is present, even in very small amounts, a white precipitate is formed. When only a very small amount of silver is present, the solution will become bluish white; when it is present in considerable amounts, the precipitate of silver chloride, AgCl, is heavy, white, and curdy. When exposed to light, the precipitate darkens because of the photo effect of silver salts in light. Silver chloride is soluble in ammonium hydroxide, differing from lead chloride, which is soluble in hot water. When a white precipitate forms during the solution of the mineral with nitric acid, the solution should be tested for both antimony and lead.

Reduction to the metal on charcoal. Silver minerals that do not contain sulfur, arsenic, antimony, bismuth, or tellurium—they are relatively few—readily produce an easily fusible globule of silver when fused with three volumes of sodium carbonate. The button of silver is bright in the blowpipe flame; after cooling, it is malleable and can be flattened with a hammer on an anvil. That the button is silver can be confirmed by the first test with the use of a chloride, as given above. When silver is present with volatile elements such as sulfur, antimony, and arsenic, a silver globule may be obtained by heating the powdered mineral alone on charcoal in the oxidizing flame. Silver per se gives no characteristic coating on charcoal. If the silver is associated with lead and antimony, a reddish to deep lilac tint will be imparted to the coating. If antimony and arsenic are present in the metallic globule, it will be dull and brittle when cold.

Sodium (Na)

Flame test. The flames of the Bunsen burner and blowpipe are colored an intense yellow by volatile sodium compounds. The test is very delicate, as only a trace of sodium can be detected.

The yellow in the sodium flame can be filtered out with a piece of blue glass with a medium color density.

This filtering action permits the reddish violet flame of potassium to be seen in the presence of sodium. The spectroscope will clearly show the presence of these elements when together.

Alkaline reaction. Soda minerals, with the exception of phosphates, silicates, and borates, become alkaline when intensely ignited before the blowpipe and will produce a brown spot when placed on moistened tumeric paper. Some minerals of other alkalies and alkali earths also produce the same reaction.

Strontium (Sr)

Flame color. Fragments of strontium minerals, when held in the forceps and ignited in the Bunsen burner flame or before the blowpipe, impart a crimson color to the flame. One may also take up the powdered mineral, moistened with hydrochloric acid, on the platinum wire and heat it in the flame to yield the color. The crimson flame of strontium should not be mistaken for that of lithium, or for the yellowish red flame of calcium. The spectroscope is most useful in determining the elements by their flame colorations.

Alkaline reaction. With the exception of the silicates and phosphates, strontium minerals become alkaline upon ignition before the blowpipe. As lithium minerals do not produce an alkaline reaction after ignition, the combination of a crimson flame and alkaline reaction is proof of the presence of strontium.

Precipitation as strontium sulfate. Strontium can be precipitated from solutions not too dilute or too acid by the addition of a few drops of dilute sulfuric acid. Upon heating, the precipitate dissolves (as distinct from lithium and calcium).

Sulfur (S)

Sulfur is found in the native state and also in compounds as sulfides and sulfates. The former are unoxidized compounds, but the latter are oxidized and require different tests.

Tests for sulfides:

Roasting in the open tube and oxidation on charcoal. The finely powdered mineral is heated in the open tube where oxidation is effected. The sulfide is decomposed, yielding the colorless gas sulfur dioxide, SO_2, detected by its sharp, pungent odor and its acid reaction. A piece of moistened blue litmus paper placed in the upper end of the open tube quickly turns red as the gas is evolved.

The finely powdered mineral may be carefully roasted on charcoal with the liberation of sulfur dioxide, which is detected by its odor. The test gives best results with those sulfides containing considerable sulfur. It is not as delicate as the preceding test.

When sulfides are roasted by holding fragments of the mineral in the forceps and heating them in the oxidizing flame of the blowpipe, the odor of sulfur dioxide will be noted. A few sulfides when ignited continue to burn in the air, because of the rapid oxidation of the abundant sulfur present.

Heating in a closed tube. A few sulfides, containing a large amount of sulfur, will part with some of their sulfur, which will condense on the walls of the tube as a sublimate. This fused sulfur is a dark amber color when hot; it changes to a pale yellow and becomes crystalline when cold. The material remaining in the tube is a lower sulfide of the metal present.

Solution in hydrochloric acid. Of the relatively few sulfides soluble in hydrochloric acid, those that are dissolved always liberate the offensive hydrogen sulfide gas, which has an odor similar to rotten eggs.

Oxidation and solution with nitric acid. When heated in concentrated nitric acid, most sulfides undergo solution or decomposition and solution, sometimes with the liberation of free sulfur, which appears as a light or dark precipitate, depending upon the impurities associated with the sulfur. This material can be removed on filter paper and tested in the closed tube. The oxidation is accompanied by the liberation of reddish vapors, the tetraoxide of nitrogen.

The filtrate is evaporated to a very small volume and to this material dilute hydrochloric acid is added. If a precipitate forms, it is removed by filtering; to this filtrate, barium chloride is added. If sulfur was present in the mineral and not entirely removed as free sulfur, a white precipitate of barium sulfate will form.

Test for sulfates:

Sulfates that are soluble in hydrochloric acid may be tested with barium chloride, which is added to a dilute hydrochloric acid solution. The resulting precipitate is of the white sulfate of barium.

Sulfates that are insoluble in hydrochloric acid must first be fused with six parts of sodium carbonate in a platinum spoon and the fusion soaked out of the spoon with water and then filtered. To the filtrate, made slightly acid with hydrochloric acid, barium chloride is added to yield a precipitate as above, if sulfur is present.

Reaction in closed tube. A few metals—aluminum, iron, and copper—form basic or hydrated sulfates that are decomposed in the closed tube, yielding a water that is strongly acid; it can be detected by inserting a piece of moistened litmus paper in the open end of the tube.

Test for either sulfides or sulfates:

The mineral is fused with three parts of sodium carbonate on a clean piece of charcoal.

A piece of the fused mineral is then placed on a clean sliver coin with a drop of water. If a sulfide or a sulfate is present, a brown spot of silver sulfide will appear on the coin. The reaction is strong and delicate. Tellurides and selenides, related chemically to sulfur, give a similar test and should be tested to be certain that the coloration is due to sulfur, although these two elements, tellurium and selenium, are relatively rare.

Tellurium (Te)

Sublimate on charcoal. When a mineral containing tellurium is heated in the oxidizing flame on charcoal, a dense white coating of tellurium oxide is formed and the flame is colored green. When the reducing flame is touched to the volatile coating, it is colored a pale green.

With bismuth flux on plaster tablet. Heating a powdered telluride with bismuth flux on a plaster tablet yields a dark brown volatile coating on the plaster.

Violet solution with concentrated sulfuric acid. When the finely powdered mineral is heated in a test tube with about 5 ml of concentrated sulfuric acid, it yields a reddish violet color if tellurium is present. When it cools, the addition of water will precipitate the tellurium as a grayish black solid with the loss of the color.

Thorium (Th)

The rather rare radioactive element thorium is usually intimately associated with the rare-earth elements; it is difficult to separate from these elements and identify, except by extensive qualitative chemical reactions.

Tin (Sn)

Reduction to metal on charcoal. An aqueous paste of a mixture consisting of the powdered mineral with two volumes of sodium carbonate is heated in the reducing flame on charcoal. The reducing action liberates the tin, which appears first in minute globules; by continued reduction they gather into larger masses of metal, which is bright under the reducing flame but becomes coated with a film during cooling. Prolonged and intense heating before the blowpipe will produce a prominent coating of white tin oxide, SnO_2, on the charcoal. The tin globules are malleable, sectile, and easily fusible.

Reduction with metallic zinc. Fragments of the mineral containing tin are placed in a large test tube with pieces of metallic zinc. Dilute hydrochloric acid is added and heated. The hydrogen, liberated from the acid by the zinc, attacks the tin-bearing mineral and reduces the surface layers to a coating of metallic tin, which may be seen with a medium-power hand lens.

Borax bead test. A borax bead is prepared on a platinum wire, and a small amount of copper oxide is added to produce a faint blue color. To this bead is added a small amount of the powdered mineral. If tin is present, the bead turns brown or ruby red.

Titanium (Ti)

Reduction with tin. The mineral is dissolved in hydrochloric acid (fused with three volumes of sodium carbonate, Na_2CO_3, if necessary—the fusion will allow the mineral to dissolve); this forms titanium tetrachloride, $TiCl_4$. Metallic tin is added to the solution and boiled. Hydrogen liberated by the tin acts to reduce the tetrachloride to the trichloride, $TiCl_3$, which gives the solution a pale violet color. It may be necessary to reduce the volume of the solution by evaporation until only a small amount of solution remains, in order to make the color visible. The usual associated elements do not interfere with this test, which is quite delicate for minerals containing more than 3 percent titanium dioxide, TiO_2.

Tungsten (W)

Salt of phosphorus bead. In the oxidizing flame, the salt of phosphorus bead containing tungsten is colorless; in the reducing flame the bead assumes a fine blue color, turning red upon the addition of ferrous sulfate, $FeSO_4$.

Reduction with metallic tin. The mineral is fused with six volumes of sodium carbonate and dissolved in hot water, as the sodium tungstate formed by fusion is water-soluble (sodium columbates are not soluble in water). After filtering to remove undissolved bases, the filtrate is acidified with hydrochloric acid, precipitating white hydrated tungstic acid, $H_2WO_4H_2O$, which is formed in the cold solution and which later becomes yellow tungstic acid, H_2WO_4. Tin is added to the tungstic acid solution and boiled, converting the tungstic acid to the tritungsten and ditungsten oxides, WO_3 and WO_2, which produce a heavy precipitate in a dark blue solution. The blue color remains after dilution with water, a distinction from columbium. Continued heating prolongs the reduction and changes the color to that of the brown ditungsten oxide, WO_2.

Uranium (U)

Salt of phosphorus bead. The oxide of uranium is soluble in the salt of phosphorus bead; in the oxidizing flame, the glass is a clear yellow but turns yellowish green on cooling. After the bead is heated in the reducing flame, the color becomes a fine green.

The colors produced in the borax bead are similar to those of iron and are therefore indecisive. The mineral is dissolved in hydrochloric acid and fused with sodium carbonate or borax if necessary (fusion will allow the mineral to be dissolved). The acid solution should be nearly neutralized with ammonium hydroxide, and powdered ammonium carbonate, $(NH_4)_2CO_3$, added. The mixture should be shaken well and let stand a few hours. Uranium is precipitated as a carbonate along with other elements that may be present, but on standing it is redissolved in the excess of ammonium carbonate and can therefore be removed in the filtrate. The filtrate is acidified with hydrochloric acid and boiled to expel the carbon dioxide. Ammonium hydroxide, added until the solution is alkaline, will precipitate uranium if it was present in the mineral. The precipitate, collected on a filter paper, will confirm the presence of uranium with a salt of phosphorus bead.

Vanadium (V)

Borax bead test. The borax bead, containing vanadium salts, is yellow in the oxidizing flame while hot and turns yellowish green to nearly colorless when cold. It is a dirty green when hot in the reducing flame, and changes to a fine green when cold.

Salt of phosphorus bead test. In the oxidizing flame, the salt of phosphorus bead is yellow to deep amber, the colors becoming weaker on cooling. In the reducing flame the bead takes on an indistinct dirty green color when hot, but changes to fine green on cooling.

Vanadium can be distinguished from chromium by the amber-colored salt of phosphorus bead when hot in the oxidizing flame; chromium produces a dirty-green-colored bead.

Precipitation with lead acetate. The powdered mineral is fused in a mixture of four parts sodium carbonate and two parts potassium nitrate in a platinum spoon. The fusion is dissolved in warm water, which removes the soluble alkali vanadates in the filtrate. The filtrate is acidified with a slight excess of acetic acid; when lead acetate is added, a precipitate of pale yellow lead vanadate results. After filtering, the presence of vanadium can be confirmed by making a salt of phosphorus bead.

Reduction with zinc. The mineral is fused in four parts sodium carbonate and two parts potassium nitrate, the fusion dissolved in sulfuric acid. Upon the addition of metallic zinc, the solution changes color from yellowish green to green, blue, and finally lavender.

Zinc (Zn)

Oxide coating on charcoal. A paste is made with water and equal volumes of the finely powdered mineral and sodium carbonate mixed with a small amount of charcoal dust. This is heated intensely in the reducing flame for about five minutes on charcoal. The zinc is reduced to the metallic form but immediately becomes volatilized and oxidized without forming a metallic globule. The oxide settles on the charcoal near the assay. It is pale yellow when hot but becomes white when cold. If the coating is deposited on an area of the charcoal previously moistened with cobalt nitrate, $Co(NO_3)_2$, the coating acquires a characteristic green color.

Green color with cobalt nitrate. When finely powdered infusible green zinc minerals are made into a paste with an aqueous solution of cobalt nitrate and intensely ignited on charcoal, they produce an assay that acquires a dark green color in the oxidizing flame.

Precipitation with sodium sulfide. The finely powdered mineral is dissolved in hydrochloric acid, after fusion with sodium carbonate if necessary. A small amount of nitric acid is added for the oxidation of iron that may be present. Ammonium hydroxide is added in excess, and the precipitate formed is removed by filtering. Sodium sulfide, $Na_2S \cdot 9H_2O$, is added to the filtrate and produces a white precipitate of zinc sulfide, if zinc is present in the mineral.

Treatment for Insoluble Silicates

As a number of the silicates are infusible and insoluble in acids, they require special treatment to permit detection of the elements by chemical or blowpipe tests.

Such minerals are first decomposed in a fusion with sodium carbonate, from which the elements can be obtained by solution with nitric acid. The quantities of materials for the fusion should be one part finely powdered mineral to three parts sodium carbonate, Na_2CO_3. The fusion can be made in a platinum spoon, on platinum foil, or by moistening the mixed ingredients and making several beads on the platinum wire. The latter has the advantage of providing fused globules, which can be crushed in the diamond mortar, then transferred to a test tube for solution in 1 : 1 nitric acid and water. The resulting solution should be evaporated to dryness with care; after cooling, several ml of hydrochloric acid are added and boiled briefly to redissolve any basic salts formed during evaporation. The silica can be removed by adding 5 ml of water, bringing the solution to boiling, and then filtering. The silica, if pure, will be white. To test for additional materials in the silica, it is well washed on the filter paper and then transferred to a test tube, by puncturing the filter paper and washing the silica from the paper. A few ml of potassium hydroxide are added and the mixture is boiled. If the silica is pure, it will go entirely into solution.

The filtrate from the silica, not including the wash from it, contains the bases. After heating the solution to boiling, a slight excess of ammonia will precipitate the ferric iron (formed by solution of the mineral in nitric acid) and aluminum as hydroxides, which can be collected on a filter paper and washed with water. If iron is present, the precipitate will be colored reddish brown and will obscure the presence of the white aluminum hydroxide. The hydroxide precipitates are transferred to a test tube with about 1 ml of water and 2 pellets of potassium hydroxide and boiled. The aluminum will go into solution, while the iron present will remain and can be removed by filtering. The filtrate is made acid with hydrochloric acid and boiled; then ammonia is added in excess to precipitate the aluminum present.

The filtrate from the iron and aluminum will contain calcium and magnesium if these were present in the mineral. The filtrate is heated to boiling, and a little ammonium oxalate added to precipitate the calcium, which appears in a very finely divided precipitate of calcium oxalate that may at first run through the filter paper. The precipitate will agglomerate if allowed to stand for ten to fifteen minutes, at which point filtration can be accomplished by repeated filtering after the pores of the paper become clogged with the precipitate. Magnesium is precipitated from the filtrate by the addition of sodium phosphate and concentrated ammonium hydroxide. When magnesium is present in small amounts, it will appear only after the solution is cooled and allowed to stand.

Acicular crystals of mesolite, a mineral in the zeolite group commonly found in the cavities of volcanic rocks, such as basalt.

GLOSSARY

Accessory minerals. Minerals present in a rock in such small amounts that they are generally disregarded when the rock is classified

Acid reaction. A positive indication of the presence of an acid, such as the turning red of blue litmus paper

Aggregate. A mixture of minerals that can be separated by mechanical means

Alums. Hydrated sulfates of two metals, such as potassium aluminum sulfate (potash alum), potassium chromium sulfate (chrome alum), or ammonium aluminum sulfate

Amorphous paramorph. A pseudomorph having the same composition as the original material—for example, calcite after aragonite; the replacing material has no definite crystalline structure and thus is considered amorphous

Amphiboles. A family of rock-forming minerals characterized by double chains of silicon tetrahedra, such as tremolite, actinolite, and hornblende

Aqua regia. A mixture of nitric and hydrochloric acids, capable of dissolving gold and platinum

Argillaceous. Shaly; characterized by the presence of clay minerals

Assay. The residue on a charcoal or plaster block following a test

Birefringence. Double refraction; the ability of a material to split light into two beams polarized in different planes

Calcareous. Containing calcium carbonate, the principal constituent of limestones

Carbonaceous. Containing carbon, usually in the form of finely divided graphite

Carbonates. A family of minerals characterized by the presence of the carbonate group (CO_3) as an essential constituent. The carbonate rocks are massive, large-scale formations composed of such minerals

Cavities. Open spaces in rocks within which crystals of minerals may form

Chlorite schist. A schist (metamorphic rock) composed chiefly of minerals of the chlorite group

Colloidal. Composed of very minute particles of colloid size, which are too small to be resolved with a normal light microscope

Cryptocrystalline. Consisting of extremely small crystals, submicroscopic in size, packed together into a dense mass, as in the agates and jaspers

Crystal face. A smooth, flat surface developed on the exterior of a crystal that formed in suitable growth conditions, without interference from neighboring crystals

Decrepitate. To crackle and break up due to roasting in a flame, such as on a charcoal block

Deflagrate. To burn with a sudden and sparkling combustion

Deposition. The settling of material, such as a vapor by sublimation; the addition of material to a growing crystal

Diaphaneity. A near-transparency caused by extremely fine texture

Dichroic. Displaying two colors when viewed in different directions. Examples: cordierite and andalusite

Dielectric. A material that does not conduct electricity, such as mica or paper

Dimorphous. Capable of existing in two crystalline forms of identical composition. Examples: calcite and aragonite (calcium carbonate); pyrite and marcasite (iron sulfide)

Diorite. A rock composed chiefly of sodic plagioclase (such as andesine), and hornblende, biotite, or pyroxene. The texture is coarse-grained

Efflorescence. Desiccation due to the loss of water spontaneously, usually accompanied by the development of a white powdery surface crust

Evolution. The production of an element due to chemical reaction; usually refers to the release of a gas, such as chlorine or oxygen

Exfoliate. To break off or peel in layers or sheets. Some minerals exfoliate when heated; rocks exfoliate due to the action of erosional agents and weathering

Fault. A break in rocks accompanied by offset of the broken pieces. Faults may be microscopic in size or hundreds of miles long

Felsite. An igneous rock with or without phenocrysts, in which the ground mass is a fine-grained aggregate of felsitic minerals, usually quartz and postassium feldspar

Ferromagnesian minerals. Minerals that contain iron and magnesium, usually dark in color

Ferruginous. Containing iron, which usually results in a red or brown coloration

Fire. Sparkles of color in a gemstone. In transparent gems, such as diamond, the phenomenon is due to dispersion, the breaking of white light into component colors of the spectrum. In the case of opal, the phenomenon is a different optical effect, known as diffraction

Fissure eruption. A volcanic eruption that bursts forth from a break in the rocks near a volcano, rather than from the mouth of a volcanic cone. Some fissure eruptions take place where no cone is actually present

Float. A piece of ore, native metal, or rock that has been separated from its parent deposit and is found in a different locality

Foliation. A laminated structure resulting from the separation of minerals into layers, usually due to metamorphic pressures

Fracture plane. A plane of irregular breaking in a mineral or rock

Friable. Gritty, easily crumbled, as in the case of a poorly cemented sandstone

Fumaroles. Holes or vents from which emerge vapors and fumes, and sometimes water, as in the case of geysers

Fusibility. The relative ease with which a mineral can be melted in a flame

Gangue. Waste rock associated with ore minerals in mining, from which the ore minerals must be separated

Gossan. A weathered "capping,"

the result of weathering, that sometimes forms over an ore body and indicates the existence of ore minerals below

Habit. The most characteristic form of a given mineral in a specific environment. For example, the typical habit for galena is cube-shaped crystals

Heavy minerals. Minerals with relatively high specific gravities that may become separated from other minerals due to the natural action of streams and tides. Concentrations of valuable materials formed by such separation are termed placers

Holocrystalline. Consisting entirely of crystallized minerals, with no glass

Hydrated. In the case of minerals, containing water as an essential constituent. A material may be hydrated, and therefore contain water, but not necessarily as an essential constituent

Hydrothermal. Formed by the action of hot-water solutions. A hydrothermal vein contains minerals deposited by moving hot waters

Inclusions. Crystals of minerals contained within another mineral. In some cases the included minerals have formed earlier and have been incorporated in the host as the latter grew. In other cases, as with rutile in corundum, the inclusions grew within the host material rather than at an earlier time

Interference. The butting of crystals up against each other, preventing the development of perfect external crystal faces

Intergrowth. The simultaneous growth of mineral crystals and consequent interference of one crystal with another

Intumescence. An enlargement, swelling, or bubbling due to heat

Isometric paramorph. A "straight" paramorph, where one crystal replaces another of the same chemical composition, the two materials having the same external shape but having different crystalline arrangements

Isomorphous replacement. A replacement of one atom for another in a solid crystalline material, to create a so-called mixed crystal

Joint system. A system of fractures in rock, usually vertical, along which usually no appreciable movement has occurred

Magma. Molten rock

Mica plate. A thin sheet of the mineral mica, of known thickness, capable of altering in a known way the colors of minerals viewed in polarized light

Native element. An element that occurs in nature chemically uncombined with other elements, such as silver, gold, or copper

Oxide. A compound usually consisting of one or more metals combined with oxygen

Oxide film. A thin film or layer that forms by the oxidation of the upper surface of a material. An example is the tarnish on silver jewelry

Paramorph. A pseudomorph having the same chemical composition as the original mineral, such as calcite after aragonite

Phenocryst. A crystal, usually with well-defined shape, occurring within a so-called ground mass, or matrix, of fine-grained crystals in an igneous rock

Placer deposit. A mass of gravel, sand, or similar material resulting from the crumbling and erosion of solid rocks and containing particles or nuggets of gold, platinum, tin, or other valuable materials that have been derived from rocks or veins

Polarized light. Light that has been forced to vibrate in a single plane perpendicular to the direction of travel of the light

Polygonal cracks. Cracks formed in soft ground when sediments lose contained water. These cracks are rarely straight, and tend to intersect in polygonal shapes and outline blocks, with as few as three or as many as eight sides

Porphyry. A rock containing conspicuous phenocrysts (well-defined crystals) within a ground mass (matrix) of fine-grained crystals in an igneous rock

Precipitate. To separate from solution, as by evaporation or by chemical reaction. A precipitate is the material so separated

Pyroxenes. A family of minerals, all silicates, characterized by the presence of "zig-zag" single chains of linked silicon tetrahedra, such as diopside, spodumene, and augite

Reagent. A substance that takes part in a chemical reaction, and is usually used to detect the presence of one or several chemical elements

Reduction. The action of separating a metal from a compound in which it is combined with oxygen, sulfur, or some other nonmetal

Reentrant angle. The angle of joining of two parts of a twinned crystal

Refractory. A heat-resisting, nonmetallic ceramic material

Residual magmatic phase. The portion of a magmatic mass that is left over after the cooling and crystallizing of the bulk of the molten material. This usually consists of a watery solution of chemical elements

Resistant minerals. Minerals that tend not to be chemically altered by weathering agents, such as water and acids produced by lichens. Such materials may be broken up physically, however, and form chemically resistant grains of sand or clay size, such as are found on beaches

Saturation. A condition in a solution of chemical elements, representing the maximum equivalent amount of a mineral that can remain in solution at given temperature and pressure conditions

Secondary minerals. Minerals formed by the dissolution and reprecipitation of minerals near the earth's surface. Such minerals usually form due to the weathering of an ore deposit and the migration of solutions to greater depths, where the dissolved material is redeposited

Sectile. Capable of being cut with a knife

Segregation. In a magma, the settling of early crystallized minerals within the still-molten rock, to form a kind of banded segregation of the early formed material at the bottom of the magma chamber

Silicates. Minerals characterized by the presence of the silicon tetrahedron, a chemically bonded unit consisting of a silicon atom tetrahedrally surrounded by four oxygen atoms

Siliceous. Containing silica (silicon oxide)

Slag. Cinders and waste material from metal refining; also used to denote partly crystalline and partly glassy material ejected in a volcanic eruption

Solution. A fluid containing dissolved material; alternately, a solid in which certain atoms have been replaced by others of a different type, to form a so-called solid solution

Specular. Consisting of minute reflective grains bonded together, to produce a "glistening" surface with many individual, brightly reflective points

Stringers. In an igneous rock, offshoots of a main intrusion that have followed small fractures and solidified, within the surrounding rocks

Sublimate. A deposit of solid material directly from the vapor state

Sublimation. The process of transition from a solid directly to the vapor state, without an intermediate liquid state, or vice versa

Subtranslucent. Almost translucent, yet allowing enough transmitted light not to be considered opaque

Sulfide. A mineral containing sulfur as a primary constituent

Syenite. A plutonic igneous rock consisting primarily of alkalic feldspar, usually with one or more dark minerals, such as hornblende or biotite

Telluride. A mineral containing tellurium as an essential constituent

Trachyte. An extrusive rock consisting essentially of alkalic feldspar and minor biotite, hornblende, or pyroxene. It is the extrusive equivalent of syenite

Transmitted light. Light that passes through a material, as opposed to the light reflected from its surfaces

Trigonal. Having three-fold symmetry

Volatility. The quality of being easily vaporized at a low temperature, as in the case of ether, alcohol, antimony, and iodine crystals

TABLE OF CHEMICAL ELEMENTS

Element	Symbol	Element	Symbol
Actinium	Ac	**Mercury**	Hg
Aluminum	Al	**Molybdenum**	Mo
Americium	Am	**Neodymium**	Nd
Antimony	Sb	**Neon**	Ne
Argon	Ar	**Neptunium**	Np
Arsenic	As	**Nickel**	Ni
Astatine	At	**Niobium**	
Barium	Ba	**(Columbium)**	Nb
Berkelium	Bk	**Nitrogen**	N
Beryllium	Be	**Nobelium**	No
Bismuth	Bi	**Osmium**	Os
Boron	B	**Oxygen**	O
Bromine	Br	**Palladium**	Pd
Cadmium	Cd	**Phosphorus**	P
Calcium	Ca	**Platinum**	Pt
Californium	Cf	**Plutonium**	Pu
Carbon	C	**Polonium**	Po
Cerium	Ce	**Potassium**	K
Cesium	Cs	**Praseodymium**	Pr
Chlorine	Cl	**Promethium**	Pm
Chromium	Cr	**Protactinium**	Pa
Cobalt	Co	**Radium**	Ra
Copper	Cu	**Radon**	Rn
Curium	Cm	**Rhenium**	Re
Dysprosium	Dy	**Rhodium**	Rh
Einsteinium	Es	**Rubidium**	Rb
Erbium	Er	**Ruthenium**	Ru
Europium	Eu	**Samarium**	Sm
Fermium	Fm	**Scandium**	Sc
Fluorine	F	**Selenium**	Se
Francium	Fr	**Silicon**	Si
Gadolinium	Gd	**Silver**	Ag
Gallium	Ga	**Sodium**	Na
Germanium	Ge	**Strontium**	Sr
Gold	Au	**Sulfur**	S
Hafnium	Hf	**Tantalum**	Ta
Helium	He	**Technetium**	Tc
Holmium	Ho	**Tellurium**	Te
Hydrogen	H	**Terbium**	Tb
Indium	In	**Thallium**	Tl
Iodine	I	**Thorium**	Th
Iridium	Ir	**Thulium**	Tm
Iron	Fe	**Tin**	Sn
Krypton	Kr	**Titanium**	Ti
Lanthanum	La	**Tungsten**	W
Lawrencium	Lr	**Uranium**	U
Lead	Pb	**Vanadium**	V
Lithium	Li	**Xenon**	Xe
Lutetium	Lu	**Ytterbium**	Yb
Magnesium	Mg	**Yttrium**	Y
Manganese	Mn	**Zinc**	Zn
Mendelevium	Md	**Zirconium**	Zr

Mineralogy and Geology Museums

UNITED STATES

ARIZONA
Arizona Mineral Museum, Phoenix
University of Arizona Mineralogical Museum, Tucson

ARKANSAS
Quartz Crystal Cave and Museum, Hot Springs
Museum of Rocks and Minerals, Murfreesboro

CALIFORNIA
San Bernardino County Museum, Bloomington
Petrified Forest, Calistoga
California Museum of Science and Industry, Los Angeles
Los Angeles County Museum of Natural History, Los Angeles
New Almaden Museum, New Almaden
Desert Museum, Randsburg
San Diego Natural History Museum, San Diego
San Diego State College Geology Museum, San Diego
California Division of Mines and Geology, Mineral Museum, San Francisco
Santa Barbara Museum of Natural History, Santa Barbara

COLORADO
Gem Village Museum, Bayfield
Denver Museum of Natural History, City Park, Denver
Colorado School of Mines Geology Museum, Golden

CONNECTICUT
Peabody Museum of Natural History, New Haven

DISTRICT OF COLUMBIA
National Museum of Natural History, Smithsonian Institution

FLORIDA
Gillespie Museum of Minerals and Monroe Heath Museum of Natural History, Stetson University, Deland

GEORGIA
Museum of Arts and Sciences, Macon

IDAHO
Craters of the Moon National Monument, Arco
Bonner County Museum, Sandpoint

ILLINOIS
Field Museum of Natural History, Chicago
Fryxell Geology Museum, Augustana College, Rock Island
Illinois State Museum, Springfield
Museum of Natural History, University of Illinois, Urbana

INDIANA
Hanover College Geology Museum, Hanover

IOWA
University of Northern Iowa Museum, Cedar Falls
Museum of History and Science, Waterloo

KANSAS
Oil Museum, Hill City

LOUISIANA
Museum of Geoscience, Louisiana State University, Baton Rouge

MAINE
Colby College Department of Geology, Waterville

MARYLAND
Maryland Academy of Sciences, Baltimore

MASSACHUSETTS
Amherst College Museum, Amherst
Harvard University Geological Museum, Cambridge
Harvard University, Mineralogical Museum of, Cambridge
Springfield Science Museum, Springfield

MICHIGAN
University of Michigan, The Exhibit Museum, Ann Arbor
Cranbrook Institute of Science, Bloomfield Hills
A. E. Seaman Mineralogical Museum, Michigan Technological University, Houghton

MINNESOTA
Minnesota Museum of Mining, Chisholm
Minneapolis Public Library, Science Museum and Planetarium, Minneapolis
Science Museum, St. Paul

MISSISSIPPI
Dunn-Seiler Museum, Mississippi State University, State College

MISSOURI
Lead Belt Mineral Museum, Flat River
Kansas City Museum of History and Science, Kansas City
Tri-State Mineral Museum, Joplin

MONTANA
Museum of the Rockies, Montana State University, Bozeman
Mineral Museum, Montana College of Mineral Science and Technology, Butte
Ft. Peck Project, Fort Peck

NEBRASKA
Chardon State College Museum, Chardon

NEVADA
Lehman Caves National Monument, Baker
Mackay School of Mines Museum, Reno

NEW JERSEY
Newark Museum, Newark
Rutgers University Geology Museum, New Brunswick
Princeton University, Museum of Natural History, Princeton

NEW MEXICO
Los Alamos County Museum, Los Alamos
Roswell Museum and Art Center, Roswell

NEW YORK
Durham Center Museum, East Durham
American Museum of Natural History, New York City
Museum of Petrified Wood, Rochester
Hudson River Museum, Yonkers

NORTH CAROLINA
Colburn Memorial Mineral Museum, Asheville
Schiele Museum of Natural History and Planetarium, Gastonia
Museum of North Carolina Minerals, Spruce Pine

OHIO
Cincinnati Museum of Natural History, Cincinnati
University of Cincinnati Geology Museum, Cincinnati
Cleveland Museum of Natural History, Cleveland
Orton Museum, Ohio State University, Columbus
Rock Circus, Amuseum, Lemoyne

OKLAHOMA
East Central State College Museum, Ada

OREGON
Oregon Museum of Science and Industry, Portland

PENNSYLVANIA
Lehigh University, Department of Geology Museum, Bethlehem
Lafayette College, Easton
Academy of Natural Sciences, Philadelphia
College of Earth and Mineral Sciences Museum, State College

SOUTH DAKOTA
Zeitner Geological Museum, Mission
Museum of Geology, South Dakota School of Mines and Technology, Rapid City

TEXAS
University of Houston, Department of Geology, Houston

UTAH
Bryce Canyon National Park, Visitor Center, Bryce Canyon
Weber State College Museum of Natural History, Ogden
Brigham Young University Earth Sciences Museum, Provo

WEST VIRGINIA
Geology Museum, Marshall University, Huntington

WISCONSIN
Neville Public Museum, Green Bay
Phetteplace Museum, Wauzeka

WYOMING
University of Wyoming Geological Museum, Laramie

CANADA

ALBERTA
Provincial Museum and Archives of Alberta, Edmonton
University of Alberta Art Gallery and Museum, Edmonton

BRITISH COLUMBIA
Geological Museum, University of British Columbia, Vancouver

NOVA SCOTIA
Miner's Museum, Glace Bay
Dalhousie University Department of Geology, Halifax

ONTARIO
Elliot Lake Nuclear and Mining Museum, Elliot Lake
Department of Geological Sciences, Queen's University, Kingston
National Mineral Collection of Canada, Ottawa
Royal Ontario Museum, Toronto

QUEBEC
Redpath Museum of McGill University, Montreal
Laval University, Musée de Mineralogie et de Géologie, Quebec

SASKATCHEWAN
Department of Geological Sciences, University of Saskatchewan, Saskatoon

Sources of Supply

LABORATORY SUPPLIES

Central Scientific Company
2600 South Kostner Avenue
Chicago, Illinois 60623

Macallister Bicknell Company
169-181 Henry Street
New Haven, Connecticut 06507

The Chemical Rubber Company
2310 Superior Avenue
Cleveland, Ohio 44114

Fisher Scientific Company
461 Riverside Avenue
Medford, Massachusetts 02155

Ward's Natural Science Establishment
P.O. Box 1712
Rochester, New York 14603
P.O. Box 1749
Monterey, California 93940

MINERAL SPECIMENS

John Ramsey's Rocks and Minerals
Sedona, Arizona 86336

Winthrop Mineral Shop
Case Road, Box 218
East Winthrop, Maine 04343

Scott's Rock Shop
1020 Mt. Rushmore Road
Custer, South Dakota 57730

Gems by Robey
Rt. 4 (hawk)
Bakersville, North Carolina 28705
June 1- October 31:
Little Switzerland, North Carolina 28749

Santa Fe Gem and Mineral Shop
3151 Cerrillos
Santa Fe, New Mexico 87501

Bentley's Minerals
1090 Matianuck Avenue
Windsor, Connecticut 06095

Lombardo Turquoise Company
Austin, Nevada 89310

Griegers Inc.
900 South Arroyo Parkway
Pasadena, California 91109

Maine Mineral Store
Trap Corner
West Paris, Maine 04289

For further sources of supply consult the buyer's guide (April issue) of *Lapidary Journal.*

Bibliography

Further Reading for the Beginner

Arem, Joel E. *Rocks and Minerals.* New York: Bantam Books, Inc., 1973. Paperback.

Dennen, William H. *Principles of Mineralogy.* New York: The Ronald Press Co., 1960.

Fenton, Carroll L. *The Rock Book.* Garden City, N.Y.: Doubleday & Co., Inc., 1970.

Holden, A., and Singer, P. *Crystals and Crystal Growing.* Garden City, N.Y.: Doubleday Anchor Books, 1960. Paperback.

Hurlbut, Cornelius S., Jr. *Dana's Manual of Mineralogy.* 18th ed. New York: John Wiley & Sons, Inc., 1971.

Hurlbut, Cornelius S., Jr. *Dana's Minerals and How to Study Them.* 3rd ed. New York: John Wiley & Sons, 1966.

Loomis, Frederic B. *Field Book of Common Rocks and Minerals.* New York: Putnam's Sons, 1956.

Mason, Brian, and Berry, L. G. *Elements of Mineralogy.* San Francisco: W. H. Freeman and Co., 1968.

Pearl, Richard M. *Rocks and Minerals.* New York: Barnes and Noble, Inc., 1956. Paperback.

Pearl, R. M. *Cleaning and Preserving Minerals.* Colorado Springs, Col.: Maxwell Publishing Co., 1971.

Pough, Frederick H. *A Field Guide to Rocks and Minerals.* Boston: Houghton Mifflin Co., 1955.

Simpson, B. *Rocks and Minerals.* Elmsford, N.Y.: Pergamon Press, Inc., 1966.

Sinkankas, John. *Mineralogy, A First Course.* New York: Van Nostrand Reinhold, 1966.

Sinkankas, John. *Gemstone and Mineral Data Book.* New York: Winchester Press, 1972.

Speckels, Milton L. *The Complete Guide to Micromounts.* Mentone, Cal.: Gembooks, 1965.

Spock, Leslie E. *Guide to the Study of Rocks.* 2nd ed. New York: Harper and Row, 1962.

Tennissen, A. C. *Colorful Mineral Identifier.* New York: Sterling Publishing Co., Inc., 1969.

Vanders, I., and Kerr, P. F. *Mineral Recognition.* New York: J. Wiley & Sons, Inc., 1967.

Wohlrabe, Raymond A. *Crystals.* Philadelphia: J. B. Lippincott Co., 1962.

Advanced Works

Bateman, Alan. *Economic Mineral Deposits.* New York: John Wiley & Sons, Inc., 1950.

Bloss, F. D. *Crystallography and Crystal Chemistry.* New York: Holt, Rinehart and Winston, Inc., 1971.

Bowen, Norman L. *The Evolution of Igneous Rocks.* New York: Dover Publications, Inc., 1956. Paperback.

Dana, E. S. *A Textbook of Mineralogy.* Edited by W. E. Ford. 4th ed. New York: John Wiley & Sons, Inc., 1932.

Dietrich, R. V. *Mineral Tables: Hand Specimen Properties of 1500 Minerals.* New York: McGraw-Hill Book Co., 1969. Paperback.

Hey, M. H. *An Index of Mineral Species and Varieties, Arranged Chemically.* Reprinted with corrections. London: British Museum (Natural History), 1962.

Kraus, Edward H., and others. *Mineralogy: An Introduction to the Study of Minerals and Crystals.* 5th ed. New York: McGraw-Hill Book Co., 1959.

Mason, Brian. *Principles of Geochemistry.* New York: John Wiley & Sons, Inc., 1958.

Przibram, Karl. *Irradiation Colors and Luminescence.* Elmsford, N.Y.: Pergamon Press, Inc., 1956.

Sinkankas, John. *Prospecting for Gemstones and Minerals.* Rev. ed. New York: Van Nostrand Reinhold, 1969.

Smith, Orsino C. *Identification and Qualitative Chemical Analysis of Minerals.* 2nd ed. New York: Van Nostrand, 1953.

Guides to Localities

Drake, H. C. *Northwest Gem Trails.* Portland, Ore.: Mineralogist Publishing Co., 1956.

Manchester, James G. *The Minerals of New York City and Its Environs.* Bulletin of the

New York Mineralogical Club. Volume 3, Number 1, 1931.
Morrill, Philip, and others. *Mineral Guide to New England.* Naples, Me.: Dillingham Natural History Museum, 1963.
Oles, Floyd and Helga. *Eastern Gem Trails.* Mentone, Cal.: Gembooks, 1967.
Simpson, Bessie W. *Gem Trails of Texas.* Glen Rose, Texas: Gem Trail Publications, 1973.
Tolsted, Laura Lu, and Swineford, Ada. *Kansas Rocks and Minerals.* 3rd ed. Lawrence, Kan.: University of Kansas, 1957.
Willman, Leon D. *Gem and Mineral Localities of the Southeastern United States.* Jacksonville, Ala.: Jacksonville State College, 1963.

Periodicals

Earth Science. P.O. Box 550, Downers Grove, Illinois 60615.
Gems and Minerals. P.O. Box 687, Mentone, California 92359.
Lapidary Journal. P.O. Box 80937, San Diego, California 92138.
Mineralogical Record. P.O. Box 783, Bowie, Maryland 20715.
Rock and Gem. Behn-Miller Publications, Inc., 16001 Ventura Boulevard, Encino, California 91316.
Rockhound. P.O. Box 328, Conroe, Texas 77301.
Rocks and Minerals. P.O Box 29, Peekskill, New York 10566.

Index

Figures in italics refer to illustrations. The page number is the page on which the caption appears.

Acknowledgments

The author wishes to extend his appreciation and thanks to the following for assistance and the loan of various materials: Geology Department, Amherst College, Amherst, Mass.; Department of Mineralogy, Harvard University, Cambridge, Mass.; American Museum of Natural History, New York, N.Y.; Field Museum of Natural History, Chicago, Ill.; Bentley's Minerals, Windsor, Conn.; N. W. Ayer and Son, New York, N.Y.; E. J. Gare and Son, Jewelers, Northampton, Mass.; Lombardo Turquoise Company, Inc., Austin, Nevada; Ramsey Rocks and Minerals, Sedona, Arizona; Schortmann's Minerals, Easthampton, Mass.; Ward's Natural Science Establishment, Rochester, N.Y.; Dr. George W. Bain, Amherst, Mass.; Raymond Black, Northampton, Mass.; Richard M. DeBowes, Sr., Metal Goods Corporation, Windsor Locks, Conn.; Charles Huber, Northampton, Mass.; Warren I. Johansson, Amherst, Mass.; R. A. Lizotte, Montague Machine Company, Turners Falls, Mass.; Dr. Arthur Montgomery, Lafayette College, Easton, Penn.; Albert B. Nelson, University of Massachusetts, Amherst, Mass.; Dr. Edmund B. Olchowski, Greenfield, Mass.; Larry W. Schoppee, Package Machinery Company, East Longmeadow, Mass.; and Harry Wayne, Raytech Industries, Stafford Springs, Conn.

Dr. Oswald Farquhar of the University of Massachusetts, Amherst, Mass., participated in the initial development of the manuscript. His work, however, prevented his active participation in the final preparation of the book.

Finally, special thanks go to my wife, Mary S. Shaub, for assistance in preparing the manuscript.